Angus

November 1980

JOURNAL
1929

ARNOLD BENNETT
JOURNAL 1929

CASSELL
& Company Limited
London Toronto Melbourne & Sydney

Limited Edition first published 1930
7/6 Edition June 1930

PRINTED IN GREAT BRITAIN

Note

Most of the entries printed here from a Journal of daily worldly things kept by me during the year 1929 bear no date. I have censored the dates, sometimes for reasons which may be apparent, sometimes for obscure reasons understood only by myself.

A. B.

JOURNAL 1929

London, New Year's Day.

I resolved not to drink cocktails any more. So in the late afternoon I went to a cocktail party. It was convened in the sacred cause of dramatic art. The party had to be serious, but also it had to be smart; it was both. Half the highbrows in London were there. I tried conscientiously to be serious about dramatic art (British variety), and succeeded fairly well at intervals. Very many cocktails were consumed.

Up to a few years ago you could not advance the cause of anything without a banquet more or less expensive. Nowadays you do it to cocktails. The change is for the better. The new method takes a shorter time, and less alcohol is swallowed. I am not in favour of cocktails; but the harm of them is exaggerated by the godly. The amount of spirituous liquid in a cocktail is trifling. The mischief is that people—especially the young—do not confine themselves to one cocktail. At this very party I was talking to a famous man. He said, while drinking a cocktail: "Cocktails are a great evil. My second daughter is nineteen to-morrow. She went out to dinner the other evening and when she came home she confessed that she had had five cocktails before dinner." I said: "But you ought to forbid it." He said: "But how can I forbid it? I do forbid it. And they come home and cheerfully announce that they have had five cocktails! You can't put your girls in prison. There it is. That's where we are to-day."

The cocktail craze will pass. And perhaps by the time it has passed we shall know the origin of the word. A cocktailed horse is, I believe, a horse which has had its tail docked. Hence its tail flounces out gaily. Hence it has an air (quite spurious) of vivacity. Hence a cocktail ought to be so-called because it gives you the jolly feeling of a horse with its tail up. But actually *is* it so-called for that reason? Nobody can say.

All I know with certainty about the cocktail is that it is a source of considerable income to caterers and bar-keepers. Even in a respectable club a cocktail costs a shilling. In a first-rate fashionable hotel it costs half-a-crown (ten francs in France). The material of a cocktail surely cannot cost more than sixpence: a liberal estimate. Add another sixpence for overhead charges—again a liberal estimate—and the hotel is left with eighteen-pence clear: a net profit of 150 per cent.

By how many inches per cocktail consumed the sacred cause of dramatic art was pushed forward at the cocktail party I cannot say. My own personal consumption of nourishment was as follows :—Olives, one. Cocktails, none.

London, January 3rd.

Royal Academy Private View of the Dutch Exhibition. I sat gazing at a superb seascape by Breughel. One of the highest authorities on the graphic arts in London came up to me. I had not seen him for a long time. Did he salute me, ask me how I was, say he was glad to see me again, or do any of the things usually done on such an occasion? No! Without any preliminaries he burst out passionately: "Don't you think Breughel is the finest painter that ever lived?" I said: "Yes, I do." "So do I," said he.

He said it, and in a way he meant it. But he would never have dared to write it. Still, I loved him for saying it. His written appreciations of art are always calm, careful, measured, judicial. Reading them, you would think that he was incapable of passion. One might even be excused for thinking that he was not a man, but a pair of scales. His impulsive exclamation to me proved that he was a man, had a beating heart, could let himself go and 'damn the consequences'. Fortunately with me he was safe.

Of course at this Private View I met lots of celebrities. What chiefly struck me about them was the likeness they bore to their caricatures. Some of them exceeded their wildest caricatures. (English caricatures are not cruel enough.)

Celebrities always deprive me of my naturalness. Either I begin to tease them or I am tongue-tied. When I met Chaliapin, on the stage of the Paris Opera, after his first performance there, I could not think of a word to say to him. Nor, apparently, could he think of a word to say to me. I just looked up at him, and he just looked down at me. Then some kindly person drew the giant away.

London, January 4th.

My only good resolution for the New Year was to read Thomas Love Peacock, one of those classical, established English authors whose reputations are invulnerable and whom nobody reads. So I got a set of his works at once, but not without difficulty. Curious that I could not get a set in a uniform cloth binding! The first bookseller whom I approached had apparently never heard of the author of "Crotchet Castle", etc. I chose "Gryll Grange" to read, and I read it in the odd moments of two days. Well, Peacock had no gift for plot; a considerable gift for the narration of an episode, but little gift for joining his episodes together. As with so many English novelists, he had not taken the trouble to learn his job. The thing is the wildest fantasy. Young hero living in a tower with seven lovely serving-maids, sisters, each of whom has a plain, bucolic, ultimately successful swain. Everything beautifully pure. An audacious modern novelist handling such a theme would certainly have got himself into trouble.

I should not be surprised if " Gryll Grange " is the most learned novel in the English language. The elderly hero, Dr. Opimian, is a great man and a greater scholar. The very numerous quotations from the Greek, Latin, French and Italian are admirably translated, and the general style of the story is admirable. The book is mature, mellow, urbane, civilized, and ironic without bitterness. I kept saying to myself: " This book is ridiculous, but ridiculous with nobility." Peacock must have been a distinguished character, if excessively odd. George Meredith married his daughter. (He ought not to have done so.) One hears that the father-in-law influenced the son-in-law. I did not see any potential Meredith in " Crotchet Castle " when I read it many years ago. But I see potential Meredith in " Gryll Grange ". Dr. Opimian is the spiritual ancestor of Meredith's Dr. Middleton, but finer—and possibly even more erudite. " Gryll Grange " is richly suffused with learning—learning carried with what elegance and what ease, displayed with what readableness! The most prodigious scholar might read it without humiliation!

London.

At lunch at the Club yesterday two University men and classical scholars, Dr. A and Mr. B, talked about the great Benjamin Jowett. They agreed heartily that Jowett's translations of both Plato and Thucydides were extremely unsatisfactory—difficulties not faced, the Greek not accurately rendered, and the style not good. Mr. B, a famous editor, said that Jowett, whom he was under at Oxford, was a very rude man. Mr. B was asked every term to a solitary lunch or a solitary dinner or both, with Jowett. Jowett would not speak. Mr. B tried to talk, of course very intimidated. Jowett would greet his remarks with such replies as: " Think again! That is a very silly remark," and so on. Mr. B's only reward for these apparently purposeless purgatories was an occasional invitation to one of Jowett's celebrity parties. Jowett was a great snob, and loved collecting celebrities. Mr. B recalled one dinner-party at which both Tennyson and Browning were present. Browning walked with him about the quad afterwards, and was most affable and interesting. Browning said to him: " They've given Tennyson a peerage, and me a Fellowship of Balliol. I prefer the Fellowship."

Jowett, it appears, wore a dress-coat (I mean in the daytime) to the end of his life. I said that this must have been one of Jowett's self-advertising stunts. Someone said not; but both Dr. A and Mr. B agreed that it was, and that the old man knew exactly what he was doing and realized the advertisement value of all his remarkably bad manners and rude if witty sayings, which were passed on from mouth to mouth. Mr. B said that the secret of Jowett's prestige was that, despite his lack of first-rate scholarship, he was a great man; also a disagreeable man. He preached sermons which were very well attended,—sermons which bore hardly on undergraduates. Jowett's chief plank was the importance to the young of planning out a career as a whole. He was always at this. He would say to undergraduates: "After you leave here you have forty years before you." Etc.

In Surrey.

I know nothing about racing. Forty years ago I saw the Derby. Or rather I didn't see the Derby; but I was on Epsom Downs on Derby Day in the midst of an innumerable crowd of persons many of whom didn't see the Derby either. Since then I have seen a country-meeting or so, a Dublin meeting, and various meetings on French race-courses, though with no interest beyond the mere general spectacle. However, to-day, in a country-house, my host, an owner of race-horses, began to talk vividly about racing people. He described the woman who, when she had won, talked to herself the whole time as she drove home, saying nothing comprehensible, just babbling and gabbling, half-unconsciously. The terrible evening gloom of those who had lost, and could afford to lose. The still more terrible gloom of those who had lost and couldn't afford to lose. He said that the betting 'system' of most of these people was to back (what they deemed to be) absolute certainties, and then by way of embroidery to make a few very long shots—50 to 1 affairs. He said that they lived for nothing but racing. They talked to all the jockeys, all the trainers, and even all the stable-boys whom they could come across.

But there was another variety of racing people —especially at Newmarket—who went in for racing simply because it brought them within ' exclusive circles '. They would have preferred to hobnob with exclusive intellectuals ; but lacking the brains necessary for the intellectual business, they courted the smart turfites, entertaining lavishly, etc., solely in order to be in ' society '. A woman present questioned the existence of this latter variety ; but my host named instance after instance, and the woman yielded. He said further that when you had ' big thing on in a big race ', the thrill during the race was unique. He named a very famous and experienced owner and said that this middle-aged cosmopolitan celebrity literally shook with excitement in such circumstances.

Later, the host took us out to see a ' promising colt ' which was loose in a field. A stableman, a shabby, horsey and entirely characteristic individual, came up to the host and broke to him the news that the promising colt had ' damaged hisself against wire '. Then we saw the colt being led back to the stable. The babyish animal limped a bit, had lost some hair in various places, and a small piece of skin from the off hind-leg. Also the animal was in a nervous state and pranced around at the slightest noise. The host was considerably upset. " Dear, dear ! This is very grave ! " he kept on murmuring to himself. He gave orders that the local vet. should be summoned instantly, and if advisable the trainer from Newmarket,—a trifle of eighty miles away. He told me the colt, six months old, was already worth £3,000. Anyhow it was a fine animal. Deep gloom!

Then we made a muddy tour of the vast field to find traces of equine hair and blood on the wire. We didn't find any. But the stablemen found them and directed us to them. There they were, plainly enough. The colt had dashed itself against the fence with such force as to bend one of the standards. You wouldn't have thought that even an infant horse could be such a fool. The host suddenly remarked: "All possessions are a mistake! All possessions are a mistake!" We returned to the house and had a melancholy tea.

Paris, February 9th.

I went to the Variétés to see the great outstanding success "Topaze". Theatre packed, and just as dirty and frowsy as when I first entered it in the year 1903. (But in those days the first four rows of the stalls were reserved to the male sex. This rule no longer obtains. The earth does revolve.) I had tickets, which were duly examined and approved by the four gentlemen in shabby evening-dress who sit at the entrance of every Paris theatre. But we couldn't get our rightful places. Two *ouvreuses* differed about the geographical situation of them. At the end of the first act we were expelled be followed by the G.W.R., the L.M.S., the L.N.E.R., the S.R., and other systems.

London.

I returned from Brighton and developed an entirely new kind of neuralgia, the fourth kind since my influenza. A man brought his niece for tea, and was witless enough to tell me that I looked tired. My latest neuralgia at once became acute. And yet doctors still begin all their treatments at the wrong end, dealing with the body first instead of with the mind first.

Blue Train.

Fog in the Channel. The ship crept forth from Dover Harbour at the rate of about a yard an hour. Three look-outs, and I made a fourth. I met various friends and acquaintances, and we all found it advisable to go below and drink champagne. (In my time I have slept on board ship for one hundred and twenty nights together, and my experience is that I prefer the worst storm to a fog at sea.) Delays apparently endless. Yet, such is the margin which the Southern Railway keeps up its sleeve, we arrived at Calais almost to time. We arrived at Paris exactly to time. The newest Blue Train is a marked improvement on its predecessors. But the railway track is no better than it was. Nor the food, whose sole merit is that it is not English. And as usual the train got later and later the further it penetrated into the slack

South. So that we reached our destination sixty minutes behind time, and for no discoverable reason except negligent sloth. And when we did arrive on the Riviera with the fabled ideal climate, everyone was full of unimaginable tales of twelve inches of snow and the destruction by frost of hundreds of thousands of trees.

Antibes.

All around the Domaine des Charmettes are vast nurseries in which flowers are grown wholesale. Tens, perhaps hundreds, of thousands of blossoms under roofs of glass, with flaps that can be raised or lowered, and straw-matting which is put on the glass-roofs or hung in front of the same. Any single blossom exhibited in a drawing-room would look beautiful and really be beautiful. But seen in the mass these blossoms lose their attractiveness by losing their individuality. They are like prisoners thronged together in an internment-camp. Difficult to believe that every man in an internment-camp has an individuality, family connections, dependants or supporters, personal ties and responsibilities!

Somewhat similar is the case of flowers in nurseries. They have no more individuality than wheat-stalks in a twenty-acre field. They are interned, subjected to rules, exposed to the sun when their governor thinks sunshine is good for them, shut into shadow when he thinks otherwise. Then they are uprooted or cut off, packed flat into boxes, despatched on long journeys in motor or train, unboxed in shops, and sold to women who stick them into vases. As soon as it finds itself in a vase, each flower resumes its individuality. It is petted and admired. It has to listen to ecstatic exclamations of joy from people who know and care nothing of its previous harsh existence, people who reck not that it may be neglected by a careless mistress, may be left to die of thirst or cold or heat, and that anyhow it will die some time, and very soon at the best, and be thrown away, chucked into a refuse-bin, amid litter and filth. Flowers have a hard life.

Juan les Pins.

This is supposed to be the most popular 'rising' resort on the Riviera. Having conscientious regard for my professional duty to see everything, good or evil, that is to be seen, I lunched here to-day. If Juan les Pins 'rises' any further, the beach, like New York, will have to rise upwards and be arranged somehow in two storeys. It is a very narrow and not very long beach, and it cannot be enlarged. The place is primarily a summer resort. Nevertheless, in April the beach was crowded with loungers, bathers, and sun-worshippers (the last deprived this day of their god). I had to pick my way on the sands among chairs and recumbent bodies. In summer it must have Margate easily beaten.

And Margate has one clear superiority over it:

Margate sands are thoroughly washed twice a day in salt water. The sands at Juan les Pins are never washed. Never! Because of the tideless sloth of the Mediterranean. I could have imagined the state of the sands in full season: but I refrained from doing so out of consideration for my own sensitive feelings. The sands of the Lido are bad enough. The sands of Juan les Pins are much worse. The domestic architecture, growing so fast that you can stand and see it grow, is shockingly bad, like all new domestic architecture from end to end of the Riviera. There must be some reason for the summer popularity of Juan les Pins. What it is I cannot conceive. The food at the big hotel was quite good. But English people do not travel twenty-four hours continuously in order to obtain quite good food. The majority of English pleasure-travellers are more interested in golf courses than in food.

Antibes, April 10th.

Motor accident. A car in front of us stopped without signalling, and we ran into its petrol-tank, which crumpled up like cardboard. An old German couple occupied the first (totally disabled) car. Tremendous Babel of English, French, Italianate-French, and German tongues. The German couple, frail and pathetic, were ready to weep amid the odour of spilled petrol. I took them into my car, and we soothed them and undertook to drive them to Mentone, where all Germans stay and always have stayed. I feared the French police, with whom one can do nothing—except in the north of France. The French chauffeur wanted to be indemnified for everything: the smashed tank, the spilled petrol, the transport of his car to Mentone, and his loss of wages while the tank was being replaced. I made the wildest promises to him, swearing upon the sacred memory of my ancestors to fulfil them, and all sorts of things. He said: "I trust you." I said: "You may."

We got away. As soon as we were beyond the boundary of the municipality in which the accident occurred, I breathed more freely. Within their own districts the police can summarily detain your car after a collision. But once outside the district of an accident neither its police nor any other can lay upon you the hand of violence. This is the only good thing I ever heard of French road-law. In due course I wrote to my insurance company, and from that moment I heard no further word of the affair.

Mediterranean Islands, July.

A party of sixteen of us had been having a long and magnificent picnic which included motor-boating from the mainland, bathing, motor-boating from one island to another, enough walking, some dozing, and a bouillabaisse the equal of which I have never eaten in Marseilles. This was the only picnic I have really enjoyed in my life! On principle I object to picnics, when they comprise a meal. I prefer a table and chairs for meals, and I think that most men not immature would agree with me. Women are different. Women prefer discomfort—up to a point. And anyhow, they have little righteousness in food. They would sooner eat bad food amid picturesqueness than good food in a prosaic room. This picnic combined picturesqueness with very good food, which food more than compensated for the unsuitability of the earth's surface as a chair or triclinium.

In the late afternoon we embarked for and reached the second island, whose buildings were limited to a café-restaurant and a monastery. We noticed that the sea was less tranquil than earlier in the day; but the island was only a couple of miles from the mainland, so that a few waves could not matter. A long dinner-table had been set for us in the open air near the somewhat primitive café-restaurant. The sun was descending. The wind was rising. The dinner was excellent, but it was eaten with a deal of sand which the wind persisted in blowing across the table, although the table was under the lee of the café-restaurant. We could hear the wind violent in the trees, and we could see the trees yielding, deferential victims of violence.

The dinner ended in deep twilight. We returned to the covered *terrasse* of the café-restaurant and drank at another long table. It grew dark. Somebody, unaware of the influence of words on the imaginative, carelessly used the word " dangerous ".

The chief organizer said lightly: "Pooh! I will telephone for the regular passenger-steamer to fetch us." But the social atmosphere was changed. Nothing, said some, could be safe in the sudden, mad Mediterranean storm. We were marooned for the night. As the oldest seafaring man present, I was invited to decide whether or not we should make the voyage. I said, "Yes, we must make it." But my decision had no effect on the minds of about half of the picnickers.

By this time there was a bridge-party, also a baccarat-party, at work, and everyone was visibly beginning to feel tired. No beds in the café-restaurant. The organizer said that he would send up to the monastery and ask the monks to lend us mattresses for the women. The proposal was not received with approval by the women. Some of them remarked with false gaiety how jolly it would be to sit up all night and play cards and watch the dawn and so on and so on. The general feeling was against departure. What interested me was the psychological explanation of the real reasons for or against departure. These were certainly not the avowed reasons. There were three real reasons. The card-players were absorbed in their games, and wished not to be disturbed. The alarmist section had passed from the normal into a highly nervous state which no rational arguments could affect. If the *Mauretania* had come to fetch us, this section would still have jibbed. A third and small section, to which I belonged, wished to depart because a programme is a programme and should be adhered to. This section insincerely laughed at the alleged risks. It would not admit the risks even to itself. It would sooner accept risks than default on a schedule. Some people are like that. Then news arrived that the steamer refused to come, on the plausible excuse that the crew had departed to their homes; but that our own motor-boat was coming. I said: "Well, if the motor-boat can safely come, it can safely go." In the high wind and in black darkness I stumbled along uneven ground, and got into a quagmire and out of it, to the little harbour. I could see the whiteness of foam. I saw the motor-boat make three attempts to enter the harbour, and fail. It succeeded at the fourth, and cheers were heard. I returned to the café-restaurant still scoffing at danger.

Then news arrived that the captain of the motor-boat had said definitely that we could not leave. Relief of the card-players! Triumph of the nervous section! Defeat of the schedule-keepers! I remained silent. I would admit nothing. In about an hour and a half, which seemed like eighteen hours or days or months, the news arrived that the captain of the motor-boat had said that the storm had fallen enough to allow us to depart in safety. All were openly relieved—for among the card-players fatigue was conquering the gambling spirit—except the nervous section. The nervous section would not trust the captain's wisdom. It spurned the argument that, assuming the captain's readiness to risk our lives, he would not risk his own. It slandered the captain, asserted that he must have had several drinks and got himself into a reckless mood and lost his judgment. But the nervous section was borne down, though it had to be dragged, protesting, to the boat.

We did depart. The extraordinary Mediterranean was calm again. And now the nervous section refused to admit that it had been wrong. Fortunately all were very hungry. About 2 a.m. we reached a night-restaurant where a nigger was saxophoning with an intensity that rendered conversation impossible. No matter. Nerves had vanished. We ate and danced. We reached our beds, some fifteen miles off, at 4.45 a.m., after escaping a motor accident by mere miracle.

London.

I went to St. Bartholomew's Hospital this afternoon to visit a relative. I cannot remember ever having been in a hospital before—except war-hospitals behind the Western Front and in London. The size of these civil hospitals is positively intimidating. St. Thomas's is a town. The London Hospital is a town. And Bart's is a town. But the largest institution I ever saw was a lunatic asylum—in Yorkshire. One is told that the London is the largest hospital. It does not seem any larger than St. Thomas's or Bart's, or as large. When one thinks that there is somebody in supreme charge of Bart's, with the whole weight of the huge organism on his or her shoulders, one is glad not to be that person.

When I go into a very large organism, I am always apprehensive of the formalities, filling up papers, submitting to cross-examination by puffed-up guardians at the gate and so on. At Bart's there were no formalities. The king of the porters gave me a brief sketch of the geography of the place. I passed through courtyards and corridors and up palatial staircases and walked direct into the right ward. I saw my relative at once, and went straight up to him; nobody challenged me, and I breathed again.

My relative told me with pride that Bart's was the oldest hospital in the world. Perhaps it is. Remarkable how inmates of an institution immediately identify themselves with the institution and take pride in it. I could never remember the name of my relative's disease, but it was something exceedingly rare and obscure. I can remember only that the cure was hydrochloric acid; so that his special diet consisted largely of liver and *pâté de foie gras* and similar odd dishes.

He is a professor of mathematics, aged sixty, a bachelor, accustomed to be absolute lord over his own domestic environment. He spoke very well of the hospital. He described to me with a sort of cynical amusement his first experience in it, including the total loss of his lordship. All his personal belongings were taken from him, except a magnificent thick dressing-gown which, perhaps because he was a professor of mathematics, he was allowed as a special favour to keep by his bed. For the rest, he had one small drawer, for books, and one spare suit of pyjamas. They say that nuns are happy. Well, this man was happy, with one small drawer and a dressing-gown for all his domain.

He had been deprived even of his will. His entire duty was to do as he was told. Rosy, full-faced, rather bald, he was as cheerful as a careless, healthy child. He related how for two days he had silently wondered why salt was not served to him with boiled eggs. At last he had discovered, by chance, that the salt was put in the egg-cup under the egg. No one had told him. No doubt the nurses had assumed that everyone would know a thing like that!

The ward was fairly large, and contained a variety of cases. I saw a small boy of 6 or 8 years, suffering from rickets, imperfect appendix, and other troubles, being led out by a nurse. There was pathos here. But to my relative rickets and troubles of the appendix were common and totally uninteresting diseases. Whereas his disease was a genuine and exclusive curiosity, and doctors came specially from distant parts of the hospitals, walked through miles of corridors, in order to inspect him. He at once became intimate with doctors.

When I left the hospital the king of the porters nodded to me as an old friend. I stood outside, and gazed at the immensity of the buildings. I walked as far round them as I could, and tried to estimate how many beds were within. And how much suffering, pain, anguish, loneliness, insomnia. The streets of the district had a strange, callous air.

London.

The managing director of a large hotel, equally expert in cookery, wine and cigars, told me at a banquet that all the true Bordeaux in the world came from a single stretch of country thirty miles long by six wide, north of Bordeaux. The quality of its grapes depended on the combined influences of general climate, sun, and soil. The same vines would not produce the same grapes if planted anywhere else in the world. It was easy for me to deduce that only a small percentage of wine described as Bordeaux really is Bordeaux. And I know that large quantities of wine produced in such *départements* as Tarn-et-Garonne are not merely sold as Bordeaux but are called Bordeaux in the district itself. The director was full of piquant information about wines. He said that Germans will pay at the rate of as much as £3 a bottle for certain Rhine wines *in the wood.* He said also, in regard to claret, that connoisseurs will differ completely as to the value of a particular wine, or year, when the price gets beyond 25*s.* a bottle. In other words, up to 25*s.* a bottle you can be sure of getting value for money, but after 25*s.* you only get individual preference for money. Whereas with German wines you can get quality as to which all connoisseurs will agree, proportionate to prices far exceeding 25*s*.

London.

A woman friend who divides her interests between England and the United States and is nearly as familiar with New York as with London, told me a very strange story. Some years ago a cocotte was strangled in New York, and a certain man, Z, was vaguely suspected of the murder. However, he went to the police, and told his story, and was believed ; and Mrs. Z stood by him, with the result that no formal accusation was brought against him. Now about this time Edward Sheldon, author of " Romance ", wrote in collaboration a play called " Lulu Bill ", which was a really terrific success, seats being sold for as high as thirty dollars apiece. It was a play about negroes, and in it a girl is strangled. Mr. Z took my friend and others to see " Lulu Bill ". Discussing the mentality of the murdered girl, Mr. Z said to my friend : " I think she *wanted* the man to strangle her. She wanted just that last thrill." At a later date Mr. Z admitted that he bought seats for " Lulu Bill " twice a week and was gradually taking all his friends to see the play.

English Seaside Hotel.

We lunched here, and had to wait half an hour because the place was packed. All the waiters were old, but decent and efficient; the head-waiter particularly; he looked like the Archbishop of Canterbury. The lunch was good and not dear. Four shillings a head.

Afterwards I inquired at the office about suites for the summer. A middle-aged dame was in charge of the reception-office. She was kindly, but marvellously *non possumus*. She assured me that *no* applications were ever, or ever could be, considered until after Easter was over. *All* applications must be in writing. *All* suites were the same: one double bedroom, one sitting-room, one bathroom. The inclusive charge for two people was twenty-six or twenty-eight guineas a week. Extra bedrooms could be had; but not adjoining a suite. This was absolutely out of the question. Single rooms would cost, inclusive, ten or twelve guineas a week in the season. *No* reduction for a baby two years old. Babies counted as adults.

The lady put every obstacle in my way. At last she said I ought to write to the manager. She was, however, in sole charge of the office, and the place was crowded with lunchers. I only made my inquiries at the office because an attractive tariff-booklet had been handed to me with the bill. The charges stated in the office were decidedly higher than those stated in the booklet.

And yet British hotel-keepers wonder why English people are so unpatriotic as to go to the continent for holidays.

Very Large Hotel in a Very Large Provincial City. I had to myself a small but officially 'double' bedroom with private bathroom. One pound per day. The bathroom, which did not need it, was centrally heated. The bedroom was not heated. I had to have a fire all the time. No bells. All orders in rooms had to be telephoned. The service, however, was efficient and worked smoothly. A writing-desk, but no waste-paper basket. When I arrived on Sunday afternoon the room felt stuffy. I tried to open the windows, but totally failed to make either of them budge. Obviously they had not been opened since they were painted, which was certainly not yesterday. I telephoned for the valet. A porter came. He also totally failed to make either window budge. He fetched another porter, and between them with immense effort they managed to open one window. The easy-chair had lost a castor. I got this remedied. I also got a w.p. basket.

The service in the grill-room, where I first dined, was not good, but the waiter was a *man* and tried. I ordered a champagne from the wine-list. After delay: "Very sorry, sir, we're out of stock." I ordered another champagne from the wine-list. After delay: "Very sorry, sir, we're out of stock." However, I was determined to have some champagne, and I succeeded at the third try. Food quite decent.

Next day I ordered a bottle of Krug, 1919. They said they had only half-bottles. So I ordered two half-bottles. Then they brought a whole bottle. This was another instance of their knowledge of their wine-cellars.

A second-rate hotel with first-rate prices. But the service in the bedrooms was on the whole exceedingly good. The food was decidedly passable, and the service thereof extremely bad. Bill for three nights £8 19*s.* 10*d.* of which £2 5*s.* was for champagne. On the other hand I had no breakfasts, and on each day I was entertained to a meal by friends.

English Seaside Hotel.

It called itself " one of the most luxurious hotels on the South Coast ". Here are some notes on the accommodation and my experiences.

No coathangers.

No luggage-stands.

No system of bell-ringing. No matter how many times I rang in the bedroom the chambermaid always came, and generally she had to fetch the waiter or the valet.

No central heating in any of the rooms.

No running water in bedrooms.

Coal fires.

Private sitting-room very large and well-furnished.

Antique private-bathroom with a lead floor under the bath.

Service slow, but not wholly evil.

No one came out from the hall-porter's office to

meet arriving cars. The hall-porter, though polite, was surly and not welcoming or even agreeable.

Boot-cleaning bad.

Clothes-brushing good.

Double-doors between rooms, and silence on the whole pretty good.

Food mediocre, except a festival dinner, which was quite good. Menus in French. Cooking in English.

By the way, although I had telephoned and written to say I wished to arrive a day earlier than expected, and had been told that that would be all right, I was not expected on arrival, and there was no fire in the bedroom as ordered. I was expected the next day. No explanation of this was given. The manager did not know until the day after my arrival that I had arrived. On the morning after my arrival I met him coming out of my sitting-room, where he had gone to see personally that everything was in order for me!

Another English Seaside Hotel.

We called in for tea. In the entrance was a card giving the times of all meals, and also: "Notice. The electric light is turned off at 11.15 p.m." Imagine it. The place was not a village, but a regular seaside town.

The tea was good.

A Hotel on the Edge of the New Forest.

Nothing but cold meat at one o'clock. Some people staying in the hotel, however, had fish. No vegetables except potatoes. No greens. No salad. Only beetroot. No fresh fruit of any kind. Yet outside was a good fruit-shop full of fruit.

London, April 23rd.

I went to see the famous 'coloured' play, "Porgy". Considered as a play it seemed to me to be negligible. Considered as a noise of wailing, singing, and praying it was far from negligible. Considered as a 'production', it was certainly very elaborate and complete; Marmoulian, the producer, is held to be in the foremost van of theatrical producing. But I could detect scarcely any modernness in the innumerable realistic detail of this production. I should call it of the school of David Belasco. I remember seeing the Belasco production of a feeble sentimental drama at the Belasco Theatre, New York. At the beginning of one act, a clock struck the hour; then, a few moments later, another clock struck the hour. "Truth to life—what!" This is what in those days in New York they called "realistic" production! If the play itself had had the tenth part of the truthfulness to life shown by the two clocks, it would have marked an epoch in the history of the stage. But the play hadn't—nor the hundredth part.

London, April 27th.

Final of the English Football Association Cup. The streets were full of charabancs carrying football enthusiasts, men and women, to see the sights of London before seeing the match. The Albert Memorial had great popularity. And indeed as a fact there are much worse architectural evils in London than the Albert Memorial. I see one rising skyward in the vicinity of Park Lane. Most of the visitors had conspicuously labelled themselves. One charabanc bore the announcement: "Reckitt's Canister Factory." I admired this esprit de corps, this industrial pride. There was something fine in it. The day, however, was remarkable to me for a reason entirely unconnected with football. Years ago I bought, together, a lovely complete edition of the "Memoirs" of St. Simon, and a lovely complete edition of the works of Rousseau, who is supposed to be responsible for the French revolution and all subsequent social progress. I had often read St. Simon, but never Rousseau. I took down the "Confessions" and began to read them. Big stuff!

I ought to have read the "Confessions" long since. They have the feel of a classic of the first order. What measured wisdom! What close and intricate psychological observation! What impartial truth! In these matters we have learned nothing new in a century and a half. If the "Confessions" were written afresh in a staccato, devil-may-care style, without verbs and commas, and signed by a young American or Irish name and dated from Paris, they would become a best-seller in two hemispheres.

London, April 29th.

" Tristan and Isolde ", conducted by Bruno Walter. Well, I prefer the bourgeois quality of the Vienna Opera to the brazen smartness of Covent Garden on a truly smart night. Not that Covent Garden was wholly smart to-night. Since we have begun to take an intelligent interest in music, Covent Garden has never been wholly smart. There is always a large admixture of genuine enthusiasts whose understanding of music surpasses their understanding of clothes. I left at the end of the second Act, not because I wasn't enjoying the performance (which was superb, except for the tenor), but because I was determined to go to bed at a reasonable hour. The megalomania of Wagner is shown as clearly in the excessive, the inhuman length of his operas as in his ruthless demands on the services, financial and other, of his friends. Only ill-balanced, one-idea'd persons can enjoy the last act of a Wagner opera as well as the first; for the reason that man's capacity for appreciation is limited. Louis of Bavaria would listen to " Tristan " twice in one night. But Louis was mad. Every first-rate performance of the best Wagner operas is attended by a number of people nearly as mad as Louis.

London, April 30th.

First night of " Paris-Bound ", an American play, put on by an American producer, Arthur Hopkins. Constantly in the press I am referred to as a regular first-nighter. In any caricature-cartoon of first-nighters I always figure. How do press-legends arise? I do not attend one first-night in six—perhaps in ten. I never go to a first-night unless I am compelled to do so for social reasons, or reasons of close friendship, or for business purposes. I hate first-nights. The performance is rarely good. And the audience is always bad, because in the main it is jaded, blasé, and thoroughly reactionary in its dramatic taste. I would sooner have a Glasgow audience than any London audience that a press-agent ever got together. As a rule a London first-night audience is hostile from the start if it suspects the dramatist of originality. This is not always true, but nearly always.

London, May 3rd.

Royal Academy Private View. As usual I saw about a hundred friends, and noted about forty frocks, but not a picture. I observed that certain celebrities went out of their way to catch the lenses of press-photographers in the courtyard. Afterwards, I suppose, they would lament to acquaintances that really it was impossible in these days to avoid publicity. Then I attended an Academy lunch in the studio of an R.A. Enjoyable; but not a word of the pictures! Perhaps this discretion was wise. Then I went to bed, exhausted. Then I got up, and took a bus and explored the furthest hinterland of Battersea, which vibrates and shimmers with romance. I saw in the window of a grocer's shop: "Relief tickets taken."

London, May 14th.

Invited to a midnight special performance of "Bull-dog Drummond" (the film) at the Tivoli, with Ronald Colman as Drummond. Samuel Goldwyn was the host. I had some people to dinner first, and took them all, and they were all in a state of considerable expectant excitement. A company of about sixty or seventy, as it were clinging together in the middle seats of the dress-circle; the rest of the immense auditorium empty. The film was stated to be the finest talkie ever done. And perhaps it was. But, as usual with films both talkie and silent, the plot was not clear; and I passed the time in repeated successful efforts to believe the impossible. Indeed the man who first said "Credo quia impossibile" must have foreseen the advent of the cinema. It is curious how a very small audience in a very large theatre is somehow afraid to applaud, and how insincere its applause sounds, even when it is not insincere. Still more curious, and disconcerting, is the appearance of a slim, elegant, fragile, spiritual-faced woman, followed by a terrific booming noise—the lady's voice! At the end we congratulated Samuel Goldwyn, who confidently predicted a very great popular success for the film. On reaching home at 1.30 a.m. I was so hungry that I had to go down into the larder and find food for myself. Thereby I learnt a lot about my own house that was previously quite unknown to me. (Later. "Bull-dog Drummond" did have a very great popular success, and the film-world seemed to me more mysterious than ever.)

London, May 27th.

My birthday. I celebrated it by going to Portland Place and undergoing what for some inexplicable reason is called a thorough 'overhaul'. I had been warned that every man over fifty ought to be 'overhauled' every few years, whether he thinks he needs it or not. Dire maladies may unobtrusively begin their awful work within you, and develop and develop quite unsuspected, and then suddenly declare open war on you, and you are dead before you are prepared for death. Moreover, had I not been suffering from chronic insomnia for many years, and must not insomnia have a cause? And so on. The advice seemed sensible. As regards insomnia, my overhauler suggested that I should take a drug, 'medinol', every night for three months or six months. Yes, such was the advice I paid for!

London, May 30th.

Day memorable to me because this evening before dinner I finished Act II of the play the idea for which had forced me to break my oath never to write another play. To the rest of the British world, however, the day was memorable as being Election Day. I went to an enormous election party in the evening, and found dozens of people seriously disturbed at the mere possibility of Labour getting a clear majority. The rancour and asperity of party politics was exposed naked in speech, tone, and gesture. Still, the food and the champagne were admirable.

London, June 4th.

As a result of betting on the election result with great press-lords, life-long electioneering experts, and sagacious persons with their fingers on the pulse of the public, I made £30 by backing my own uninstructed opinion against theirs. A small reward financially, but large in the dangerous gift of self-esteem.

London, June 5th.

I received the detailed report of my overhauler, via my ordinary doctor, in accordance with medical etiquette. There was nothing the matter with me at all. Blood pressure right. Heart very fine. Lungs very fine. Arteries suitable to the age of 32 instead of 62. The liver somewhat less brilliant than the other parts of the body, but still an excellent specimen of this great organ. No light thrown on my chronic insomnia. As a fact, another doctor, not professionally consulted, had once uttered to me the final word about my insomnia: "It's simply this," he said. "You happen to be a bad sleeper." I have never got beyond that!

London, June 6th.

The Election has had a very disconcerting effect on the more 'sporting' of my clubs. In 1924, in similar circumstances, lots of men were saying that they would leave the country and settle abroad. They also said: "What we want in this country is a Mussolini." Mussolini's ears must have burned. This time, less is being said about leaving the country; but the urgent cry is still for a Mussolini, and Mussolini is still held to be the greatest European statesman since—Disraeli.

London, June 7th.

To the Victoria Palace to see "The Show's the Thing", with Gracie Fields therein. Her husband, Archie Pitt, appears to be the author of this musical piece. My admiration for the show which is the thing has reserves. But my admiration for the way in which the author infallibly hits the nail on the head every time is without reserve. Archie Pitt knows the taste of his audience profoundly. Loud sallies of laughter going off like a gun every thirty seconds or so—throughout the evening. This kind of a success proves by comparison how chill and partial is the enthusiasm aroused by successful pieces in an ordinary West End theatre. Certainly it is a very agreeable sensation to see two thousand people unanimous in uncontrolled mirth.

London, June 18th.

I finished the play, except the title, for which I have not yet been visited with a single useful idea. Of course I am temporarily convinced that the play is very original and very brilliant. But this agreeable impression will not last. I tell my friends that I still have two plays unproduced—one written sixteen years ago—and that this new play will never be produced. Also that the offer of the entire earth would not induce me ever to write another play. I will not harrow my own feelings by recalling my frightful trials on the West End stage since the production of " Judith " (the play, not the opera).

London, June 25th.

Another oath broken. After some fantastic experiences at my own first-nights, I had sworn never to attend another. But when I told the authorities of Covent Garden that I should not be present at the first performance of the Goossens-Me opera " Judith ", there was such horrified, outraged protest that I accepted a box on the spot. And there I sat on the first-night, hiding behind a curtain, and surveying the crowded house. My highly nervous state was mitigated by the realization of the unquestionable fact that I was not Eugene Goossens, exposed defenceless to the public and conducting the orchestra. I kept carefully in the box, but well-intentioned friends and quidnuncs would insist on visiting me both before and after the performance. I had not the courage to tell them that, with the important exception of loud and prolonged applause, all an author wants on a first-night is to be left alone. I paid what I was afraid would be a state visit to the prima donna.

But it was not in the least stately. After I had kissed her hand we forgot ceremony and were realistic with one another about all manner of things, and laughed like girl and boy.

Portsmouth.

I came down here for the first performance of a play 'prior to London production'. Would that theatres in the West End of London were half as picturesque, charming and commodious as the Theatre Royal, Portsmouth. There isn't a theatre in the West End, except Covent Garden, to compare with it. The cast was pure West End. But Portsmouth audiences are not to be intimidated. They are constantly seeing genuine West End productions. At the rival theatre (no, not the rival, because both theatres are under one proprietorship) a first-class London star was 'opening' on that self-same night with a new play. The hotel was markedly theatrical. After the performance groups of actors, actresses and playwrights supped endlessly in its dining-room, whose waiters are accustomed to late hours. I talked with various artists.

The talk was solely, endlessly, of the stage. I doubt whether actors and actresses ever talk about anything but the stage. I am quite sure they never tire of the subject. As an inexhaustible subject, the stage beats hunting, golf and fishing hollow. At a few minutes to 1 a.m. I bade good-night. Everybody seemed surprised that I should retire so early. Everybody was still eating and drinking and talking eagerly, and without the least sign of fatigue, about the stage. And the waiters were still quite lively. The night-porter told me how stars, whom he named by name, would stay up talking till 5 a.m. He thought nothing of that. Actors and actresses must have the physique of explorers. I can see no reason why they should ever die. Some of them never do. I met a star who had been a star of the first magnitude when I was a youth aged twenty. She was as vigorous, impulsive and capricious as ever.

London.

From the first chapter of Sir Cecil Walsh's "The Agra Double-Murder Case": "George Eliot somewhere says that women would be spared half their sufferings if they would only refrain from saying what they are prompted to utter, when they know before they utter it that it would be better left unsaid." George Eliot may not have been the greatest novelist that ever lived, but at any rate no woman, probably, was ever more free from sex-bias than she. This freedom must have impaired her feminine charm.

London.

Men of business are baffling creatures. To-day I invited to lunch at a grill-room a prominent business man, because I wanted to straighten out certain obscure matters. A man accustomed to think in millions—other people's millions! He arrived a few minutes late, saying that he was excessively busy and had enough work to occupy two men. We drank cocktails. We then lunched—excellently. From time to time he repeated that he was an excessively busy man. We talked small-talk, about everything except our joint business. Again and again I led him gently to the brink of our business, but he could not persuade himself to plunge. Possibly the water looked too chilly. At three o'clock we were still conversing at large. At three-fifteen I said to myself: "When does this fellow intend to go?" At three-thirty I said to myself: "This fellow will just *have* to go soon." At nearly four o'clock I said to him: "I'm very sorry, but I must go." "So must I," said he. He went. And I doubt not that during the rest of the day and during all succeeding days, he would be assuring people that he was an excessively busy man. As for me, I had wasted two hours and a half of life's brief span, and about a couple of pounds.

London.

Dinner and interview with a star film-director. I was told that he was the finest and the most successful film-director in the country. I had never seen anything of his on the screen. A youngish man, with a clear, penetrating voice, trained no doubt to make itself heard in the immensities of studios, but more than adequate for a dinner-table. I had written and sold a film scenario, and at the request of the purchasing firm the director was considering the same.

At first he assumed the air of a puissant lawgiver. When resolutely tackled, however, he changed the air for another one, and we became almost equals. He argued on the following lines: "Now the hero of your story is a financier. Now would a street-barrow woman in Hoxton understand about getting an overdraft at a Bank? That is the test, the street-barrow woman in Hoxton. Is office work and typewriting romantic? *Is* it? Now if we could have the story lowered in class, if for instance you could make the financier a ringmaster in a circus, now that would be colourful. What we want is colour." And so on in similar style.

I argued him out of every point, and soon his principal phrase was, "Oh! I agree." But though he agreed point by point, he did not agree in the least on the whole question. I yielded on nothing He yielded on everything, and said at last he would think the matter over. But I was quite certain that he never would think it over.

The purchasing firm had specially commissioned me to write a story that did not resemble the ordinary film-story. They knew my more notorious books, and desired a story in the manner and on the plane of those books. I had asked them if they wanted truth to life, and they had replied that truth to life was precisely what they did want and that on the screen truth to life was coming more and more into fashion! Audiences were getting tired of sobstuff, etc. I now perceived that, if audiences were getting tired of sobstuff, film-directors were not. Film-directors still had their eye on the purely imaginary woman in Hoxton. The purchasing firm had reckoned without their film-directors, and film-directors had the last word.

This particular film-director could choose between film-stories. He was very far from being a fool. Intelligent in his own way; but it was a crude way. There was certainly something of the artist in him. Some creative fire in him. I liked him.

When we had thrashed the story all to bits and neither he nor I had anything more to say about it, he grew communicative about himself, in reply to my questions, and revealed himself a regular figure out of film-land. He said: "I can never begin work until about eleven-thirty in the morning. I have a glass of sherry then, and that starts the flow of ideas. You must have the flow. The film must move rapidly, and so your ideas must come rapidly." An odd argument, but I don't think he could see any flaw in it.

I found that in order to impress him I had to boast. Hence I did boast. I called down rather sharply one of the head-waiters who had been too curt with me. This episode obviously increased the director's respect for me. When the interview was over he said he should not require his car, and amiably offered to send me home in it. It was a magnificent, a glorious car, the car of the legendary film-director. I thanked him and said that I might just as well go home in my own.

London.

I went up to Elstree to-day to lunch with a film-manager and another film-producer, and to talk business. I had been commissioned to write a silent film and had written and delivered it, and received a substantial payment on account of it. Then the talkies came along and blew out the fashion for silent films as a man blows out a match. So that I am desired to transform my film from a silent into a talkie. Much discussion and no apparent result. It is astonishing, the rapidity with which a fashion can die. And yet, is it astonishing? The manager told me that at a certain large cinema theatre "The Singing Fool" had raised the receipts from £2,000 to £6,000 a week.

London.

The longest lunch I ever attended lasted from 2 p.m. till 7.20 p.m. This was in Dublin in 1917. I had another long lunch to-day, at home. It lasted from 1.30 to 5.30. The visitors were the film-magnates. Magnificent discussions; illimitable goodwill; innumerable compliments on both sides; but we could not agree. Therefore no result.

London.

To a birthday party to-night two well-known English male songsters had been brought from a

Paris cabaret show in order to entertain us. Said to be a brilliant success in Paris. But we talked ruthlessly through their first turn. When I say 'we' I mean most of us. I felt rather sorry for them, and I went into another room to chat with them while they were waiting their next turn. My sorrow was misplaced. I ought to have felt sorry for the interrupters rather than for the singers. The singers, safely encased in an armour of self-complacency, had no need whatever of sympathy. I am always interested in the question of sleep. I asked them what time they managed to go to bed. One of them said: "Generally about half-past four. But we get good sleep, and as we are always travelling about we have learned to sleep anywhere, in any position." I must say that they had the fresh, healthy innocent complexions of small boys who are regularly put to bed and tucked up by their mothers at 9 p.m. Nobody could possibly have guessed that they were cabaret performers.

London.

Authentic revelations about the central administration of justice in Ireland before Ireland, or part of her, became a Free State. A Lord Justice of Appeal had to wear on state occasions a court dress with a train. One Lord Justice named to me bought his dress third-hand for £50 from a second Lord Justice who had bought it from a third Lord Justice, its original owner, who had paid 250 guineas for it. The dress was destroyed in the destruction of the Four Courts by the rebels. The Government asked Lord Justice No. 1 what compensation he wanted. He asked for £200 and got it. A train needs a train-bearer. Lord Justice No. 1 did the usual thing: he appointed a relative to be his train-bearer. This boy had to bear the train four times a year for about half an hour each time. Salary £100 per annum.

Apropos, my informant told me that in the glorious old days of the Exchequer Division the judges thereof had the patronage of various sinecures (£1,000 to £1,500 a year each) and that they invariably appointed their fellow-judges to these sinecures. And my informant also told me that the Irish Attorney-General, before the rule of Sinn Fein, had to see all indictments at a fee of three guineas apiece, and that his income from this source had been known to run as high as £25,000 in a single year. I received other similar information, but nobody would believe it in print. I have allowed it to slip my memory. Some day perhaps I shall obtain equally startling stuff about the financial idiosyncrasies of Sinn Fein.

London.

A theatrical producer came in to see me this afternoon. He told me some of his marvellous adventures as producer of an 'All British' musical comedy, an enterprise controlled by a group of capitalists with plenty of capital but no knowledge of the stage and no knowledge of their own minds. In other words, they had money to burn but could never decide just how to burn it. They were always talking about 'extra numbers'—that leading feature of so many successful or semi-successful musical comedies. But they could never find lyric-writers for themselves, nor definitely engage a lyric-writer when one had been found for them. One thing was once decided, namely, that as the plot consisted of a Canadian love-story, a big spectacular forest scene, complete with lyrics and choruses of maidens and Red Indians, etc., would be indispensable in the second Act.

The great all-British musical comedy was being rehearsed, concocted, and built up in the provinces. My producer was therefore instructed to go to London and spend up to £1,500 on this single extra number and its costumes. He was just setting off when he said to one of the capitalists: "But what about the extra number itself? Have you got it?" The capitalist replied: "No, we haven't got it yet. I suppose we ought to get it, oughtn't we?" My producer said: "I think it would perhaps be safer to get the number first, so that we can know what the words and music are before ordering the 'set' and the dresses." The capitalists concurred. But nothing was done; nothing was ordered; and the extra number was neither written nor commissioned. The matter merely dropped. And so on and so on.

London.

I lunched with an old Liberal politician and University big-wig. He was full of courtesy, detailed reminiscences, and general ironic wisdom. But he was absolutely sure that the Conservatives would have a clear majority in the next House. Nothing would move him from this position. He told us he had laid a bet of a silk hat on it. He told us that, in connection with some electioneering affair, and long ago, he had stood Lord Rosebery a lunch. He had asked Lord Rosebery what he would drink. Rosebery said: "What are *your* ideas on the subject?" My friend said: "Well, I fought one election on stout, and lost it. Then I fought another election on champagne, and won it." Rosebery said: "We must have champagne then." They did.

London.

A new experience to-day. Eating Caucasian bear. Never had I dared to hope for the flavour of Caucasian bear in my mouth. The restaurateur, who had given me several days' notice of the arrival of the delicacy from the Caucasus, told me that he was the first restaurateur in London to provide it. This was the fourth time he had managed to obtain it. How Caucasian bear is got out of the Caucasus and transported to London I don't know, and I omitted to ask.

The bear, or an edible part of it, is soaked in Burgundy for one week, and then braised—well done. It is served with two sauces poured on it together—one brown, one yellow; and with mashed chestnuts. I found it *très fin* and entirely delectable. The flesh does not taste like any other flesh. The taste nearest to it is that of venison, but it is better than venison. I ate venison in clubs in the war, when beef and mutton could not be had, and venison made little appeal to my palate. It could be eaten. But horse can be eaten, and I have eaten it. Horse is rather like third-rate beef, sweetened with sugar.

The restaurateur invited me to have a glass of vodka. Now vodka was quite unknown to me. I ordered a glass. He produced an enormous bottle, with double eagles on it, and of the dimensions of a jeroboam. He said he had bought three such bottles from a Russian prince in Paris. Twenty pounds apiece, and cheap. It was a grand drink.

Italy.

An English bank is inhuman or godlike. Its attitude towards ordinary customers implies that it is conferring a favour upon them by doing business with them at all and that they ought to consider themselves indeed a fortunate lot. It cannot or will not recognize that it is a mere shop for the sale of monetary facilities, and that there are rival shops. That it exists for the convenience of its customers, and not vice versa, is an idea which apparently seldom occurs to it. Sometimes, however, by firmness, the recognition of this important fact can be obtained. I once desired to carry out a perfectly usual transaction with a bank. The manager asked me to go and see him. I refused. He came to see me. He permitted himself to remark that I was living in a rather large house. I permitted myself to suggest that he was just possibly getting away from the point. The transaction was executed immediately. And I have often heard tales from customers who knew exactly what they wanted and exactly how much they were prepared to pay for it, and who succeeded in rendering bank-managers human. It is fair to say that English banks are as a rule business-like and rapid in their methods.

Now French and Italian banks are human. They are very human. I have done plenty of business with French and Italian banks. Yesterday I went into a typical large branch of an Italian bank; and everything happened according to precedent. Italian banks close from 12 to 2. Such a system would not work in England, but it works smoothly enough in both Italy and France. At ten minutes past two the numerous staff comes strolling casually in from its lunch, smoking cigarettes. Italian bank-clerks seem to be unable to do business without tobacco. And why should they do business without tobacco? Tobacco is humanizing. Their manners are exquisite; their charm is notable.

At the first guess I went to the right counter, behind which some half-dozen clerks were more or less busy in a cubicle. I presented a cheque—not a foreign cheque, but one of the Bank's own cheques. I furnished evidence of identity. I was most urbanely received. The entire half-dozen young males showed a friendly interest in me. Then I said: "I want this money in English sterling."

"*Sterlina!*" exclaimed the youth attending to me, astounded. "Ah! We must go upstairs."

He escorted me upstairs to another and a vaster room. He telephoned to the cashier downstairs: "Have we fifty pounds in sterling?" A pause. The clerk smiled. Yes, the Bank had in its coffers fifty pounds in sterling. Then began the filling up of forms, with carbon duplicates. A tremendous affair. My full name, the name of my father, my permanent domicile, where I had come from, where I was staying. Then I started to endorse the cheque, which required two separate signatures. With perfect tact, the clerk stopped me.

"Excuse me," said he. "The signature on the front of the cheque ought to have been written before you came into the Bank."

"But I will sign it in your presence," said I.

"Ah, sir! We have our rules." Then followed a long palaver among the staff.

"I must consult the Director," said he.

He departed to consult the Director, who presently arrived to see me. The manners of the Director were marvellous: a lesson to all Britons. They were comparable to those of the late Lord Chaplin, whose social deportment I have never seen equalled—in England. The Director agreed with me that I might perform the first signature in the Bank itself, and left me with an enchanting bow and smile. Then the second signature. Then more forms which had to be signed. The clerks were in continual consultation as to procedure.

"Now we will go down to the cashier," said my special clerk. "I will accompany you."

We descended to the cashier. A third signature was demanded. In all, I wrote twelve signatures. But I got the sterling. I also got humaneness, charm, and courtesy. Also my affair had engaged the attention of eleven clerks and the supreme Director. At the close everybody appeared to be very pleased, and relieved. On everybody's face was an indication that a miracle had been accomplished. I myself felt that a miracle had been accomplished. True, it had taken thirty-three minutes: but a miracle it was. I went forth into the hot blinding sunshine of nearly three o'clock. All this happened in an illustrious city where tourists are as common as flies; and in my view it was quite as interesting as any of the city's storied monuments.

Lago di Garda.

The hotel-pension is quite the best I ever stayed in. A former Archbishop of Canterbury is reported to have said, as to food: "All I ask for is a plain but perfect table." We had what the Archbishop of simple tastes asked for. And everything else about the place was perfect, including the service, which was rather super-perfect, for the reason that, having arrived long before the hullabaloo of the summer season had set in, we had the entire hotel to ourselves.

The establishment had been built and founded, and was run, by an Englishman Italianate, who was the chef. The place has only two disadvantages. The steep, terraced garden, being very young, offers no shade—indeed after 11 a.m. there is no shade outdoors within half a mile. And the insects (not mosquitoes, however), which are restless and inquisitive, and formidable in appearance if not in size. Had this size been multiplied by ten the insects would have emptied the district of human beings and ruled a kingdom.

The hillside above, like the garden, is as steep as a roof. The faint sound of a bell descends from its hidden verdurous height. I drive up. It is a severe climb, even for a car. An interminable succession of acute hair-pin bends. I reach at last a village, San Zeno di Montagna. The altitude is marked on a wall, 700 metres odd—appreciably over two thousand feet. Every Italian village has five profane phenomena: a cobbler's, a barber's, a general store, an unkempt café or so, and a municipal decree written on paper, signed by his Worship the Mayor, and affixed to the façade of a prominent house. In other respects San Zeno di Montagna seems to be uncorrupted by the latest civilization, save for a young woman or two nicely shod and with coiffures whose elaboration has cost time if not money. The church goes without saying. It holds nothing to astonish, unless you are capable of being astonished by a large portrait of San Zeno himself in modern stained glass. A workman has taken out one of the windows and, perched on the sill, is working with a noisy hammer thereat. The village priest, under an umbrella, promenades to and fro, keeping a prelatical eye upon him. A few indifferent inhabitants, no doubt the oldest or the fattest, move slowly around the slatternly, dusty, half-ruinous hamlet on their vocations agricultural or commercial. The two cafés are empty.

The great fact is the sun overhead, much too bright to be visible, shooting down cataracts of hottest light. Millions of insects unseen whirr and creak, as though their machinery needed lubrication. I look through the boughs of an untidy orchard, and see the changing sheen of the vast surface of Lake Garda beneath me, the largest and the least spoilt of the North Italian lakes. Over thirty miles long; ten miles at its broadest and a couple of miles at its narrowest. North, west and east it is shut in by mountains, quite a number of them rising to seven thousand feet. And at the foot of these thirty or more villages. Several of them towns, with dubious imitations of grand hotels, white tram-cars, Kodak shops, and emporia for the sale of English and German newspapers. The population of one town attains five thousand. And half-way up the ranges lie large villages, with no hotels, no nothing except cobbler's, cafés, barber's, small stores, and mayoral ukases. Perilous roads hang like cotton threads on the slopes; and five miles of them are as one inch.

And rising very high above all this, range behind range, the naked summits of the coloured mountains, never climbed, never approached, utterly immaculate. An immense quantity of the earth's surface in Italy is wasted by nature, serves no purpose save to impress the unaccustomed eye. Impress the unaccustomed eye it certainly does. The scenery is lovely, majestic, the scenery of a dream or of a picture-postcard. The eye can discover in the distance naught but a dreamy, hazy, ideal beauty conceived on a terrifying scale. If beauty and terror are the desiderata of the artist, here he has them, incredibly combined.

The lake is empty. Not a steamer in view, not a barge in view. Only one islet (but there are two still tinier islets out of sight to the north). This is called the island of Garda. I saw it close yesterday. It belongs to the Borghese family. Of course it would—or to the Scaliger family. It has a stucco palace on it in the worst taste, and a whole little street of appurtenant out-houses in better taste. Apparently the palace is uninhabited; but you never know. It is not "open to the public". The leases of some historic Italian palaces contain a clause compelling the tenant to deliver his drawing-rooms one day a week to the common gaze—for a trifling fee. But the Borghese family are superior to such rules. The island has a queer reputation. Not so very many years ago a Borghese princess fell off a terrace parapet while watering flowers and was drowned in the sea of the lake. People spoke of suicide. Soon afterwards the next tenants, an amorous pair, were drowned out of a boat in a squall. Thereupon the rumour of a Borghese suicide suddenly became an accepted truth. Nobody could deny that the unhappy princess did not fall off the parapet, but threw herself off in despair. And the island was accursed, fatal to human life. There the island lies, a glistening spot hung between the sea of sunshine and the sea of the lake.

San Vigilio.

After dinner in the evening heat of the hotel gravelled garden, which gives on to Lake Garda and on to a private chapel whose doors open only twice a year, while we were smoking and drinking iced lemon-squashes or gin-fizzes, we were told that there was "a guitar at the other place". We went at once to the other place, which is a public-house or café provided by the hotel for our social inferiors. It lies on the harbour. We were furnished with chairs on the opposite side of the harbour; but as the port is at most a hundred feet across, the view and the acoustics were perfect. Twenty or thirty social inferiors sitting al fresco at tables, and smoking, drinking, and chattering. The lights of the interior of the café illuminated them. People standing here and there. A girl or two in white standing, in postures perhaps more studied than they would admit. In a doorway a tall and handsome young man in shirt and trousers bent lovingly over yet another girl in white; they were talking vivaciously, but in murmurs. Behind the central seated group sat the guitarist. White boats and a laden barge in the tiny harbour.

A man suddenly stood up. Silence. In a baritone voice he sang a fairly good song; the guitarist seemed to be reducing the accompaniment to the lowest minimum. Well, perhaps it was not much of a song; nor was the voice very powerful or brilliant; but the singer sang agreeably, with taste, without exaggeration, artistically. Quiet applause. Then from the swell part of the hotel a gramophone started up, and gave a great rendering of a tenor-soprano duet by Verdi. The tenor had a magnificent professional voice, conscious of its terrific strength. Very different from the small voice of the man who had sung to the guitar. But there was a sort of over-accomplished insolence in the great professional voice; and in the humility of the surroundings and the meditative heat of the night, one somehow preferred the amateur's. Then the amateur sang again, a worse song than before, but with the same artistic discretion. Frogs in the water croaked like ducks in opposition, and a youth threw stones to quieten them, and failed. Three muttering men at the end of the south mole of the harbour argued themselves nearly into a quarrel.

Two of them walked back sneering to the café; the third remained solitary, spitting into the lake.

The singer had not allowed himself to be disturbed. He sang on, and quietly finished; the guitar twanged a full-close. More quiet applause. Then the guitarist himself, still seated, gently sang a song, and got the chief applause of the evening. The concert was finished. I heard a lapping behind me. Lo! A little water-gate, admitting water under the hotel! A mysterious lapping, emerging from the mystery of the piled foundations of the fifteenth-century house.

We got into a white dinghy and rowed slowly out into the lake. All round in the distances flickered the lights of townlets and villages. Not a breath in the air; not a ripple on the water. Utter stillness. The heat seemed to descend from above. You were conscious of your pleasantly perspiring body, more conscious of your body than a man could ever be in England. (With women, who wear relatively so little, it is different.) Then a scarcely noticeable stir in the air. Not a breeze, not even a zephyr, but a faint general disturbance. Then a zephyr; then a breeze, in a definite direction. It is the night-wind that regularly arrives over the lake from somewhere.

Then a shape, or the semblance of a shape, ghostly, afar off. It is something; it is an illusion; it is something. Then it is a moving boat. Then it is a boat with men in it and a sail. It appears to be bearing down upon us, gigantic. It is a large, white open boat with one of those queer rhomboidal sails, the boom of which is placed high enough on the mast to enable men to stand upright under it and row. It approaches, out of vagueness and nonentity, and becomes a sharp reality. Voices of the men in it. It swishes by at a great pace, swerves, and makes the little harbour in first-class style. The lake is now an expanse of ripples, and the wind, penetrating your thin clothes, cools you.

San Vigilio.

A red flag, hung at the end of the south mole of the doll's harbour of San Vigilio, warned the steamer to stop. She stopped outside the harbour; if she had tried to enter it, small though she was, she would have either burst it or buckled up. We went on board. Not a very clean steamer, but we had her practically to ourselves. The most important other passenger was a cow, tethered to the stem. No formalities about tickets. A man who might have been a sailor or a milkman strolled along presently, and asked whether we would like tickets. I said we would. I said also that we would like the awning spread, this being the hottest part of the day, and the downpour of heat and shine being tremendous. The awning was at once spread. I also said I should like some coffee. The surprising thing was, not that the coffee should be supplied, but that it should be drinkable. Indeed the coffee was excellent.

The steamer had already called at Torri del Benaco, and was making the first of its numerous crossings of the lake. The small harbour of Torri is somewhat less small than that of San Vigilio. For Torri is quite a town, with an arcaded hotel, and bright girls sitting under sunshades, and a street of shops, and a camera-store and all sorts of things. A sizeable schooner-barge with lemon-tinted sails lay in the port, in addition to row-boats. Women were washing clothes near by in full sun-shine. The cow went ashore with much nonchalant dignity. . . .

Now we were in the middle of the lake heading for the western or fashionable side, which in places calls itself a Riviera. Maderno, the next call, is even more important than Torri, with a more important harbour, white tram-cars, two churches, and a vendor of coloured but insipid ices. Here we had a shock. We were instructed to change steamers. We had almost owned the first steamer. Aboard the second one, which was very full of Teutons (probably Austrians), we were nobodies. We could not find a single seat at a deck-table. We had to squeeze in anywhere on the side-benches.

But we had a magnificent view of a Teutonic family of three—father, mother, and daughter—eating an enormous lunch to a succession of bottles of beer. It was then 3.30. At 5.30 the lunch was not finished, nor the succession of beer-bottles ended. Father stood up and stretched himself. He was a colossus, and we admitted that a father so enormous was bound to eat continuously. Further, his relations with his women were charming.

Next, Bogliaco, still on the fashionable side. The usual harbour, the usual church, the usual campanile, the usual many-tinted picturesqueness. The prevalence of harbours, by the way, is good proof that Lake Garda suffers from sudden and severe storms. There are more harbours on Garda's margin than on all the south coast of England. But then Garda is entitled to behave like a sea. It is half as long again as the distance between Dover and Calais; and its mountains, far higher than anything in England or even in Scotland, can have no tranquillizing influence on the waters which they hem in. Squalls must come down these steep slopes like thunderbolts. A newspaper-seller came on board at Bogliaco, and every newspaper he carried was either German or Austrian. With him, a woman offering baskets of lemons and the foliage thereof.

We were getting into the lemon country. I heard that the lemon industry was as depressed as British agriculture; but there were plenty of plantations, with row upon row of palisades upon which is stretched in winter some kind of protection for the delicate plants.

So the steamer went on crossing and re-crossing, the scenery growing wilder and more rugged at every slant northwards. At Tignale only a bed disembarked; it was met by its owner. Tignale appeared to consist of one large building, a long, thin waterfall, and a hundred lemon-plantations. The cliffs now rose perpendicular out of the water, very high and formidable and intimidating. The steamer paddled right under them. . . . All the softness characteristic of the southern end of the lake had vanished, as though we had travelled from Naples to Bergen. But the lake and its mountains still shimmered in haze and sun.

Malcesina, on the unfashionable side, with a castle in addition to a campanile, was the most picturesque and warmly coloured of all the townlets. The parasol-tents and the tablecloths in a garden café there had lessons in them for the lady-decorators of Beauchamp Place and similar streets in London. On a wall I read the curious words: "Viva Mazzini." Clearly Malcesina had not yet caught up with the Duce. Another crossing. Limone. A townlet built into the very face of a terrific beetling cliff. The back wall of the hotel must have been the cliff itself. No sunlight here after 5.30 p.m. at the latest.

At Torbola there was a Grand Hotel, with uniformed porter, a car, and a bus, all highly civilized. The lake narrows. The mountains take on that violence of picturesqueness which demands to be photographed. Then at last, after about three hours' voyaging, Riva, the head of the lake. Until the war Riva was Austrian. Riva is a real town. Its picturesqueness, with that of the mountains behind it, is extreme. A likeable town, full of surprises and variety. It has all the resources of civilization, except traffic-directing policemen. We bought an English newspaper in Riva, only to discover later that the sheet was eight days old. Cocktails in Riva. Also touts. And shops for the sale of everything. And we felt as if we had been furrowing the high seas for a week and had arrived in a foreign land.

Riva.

We had counted on a return steamer as far as Malcesina, at which piece of picturesqueness we had arranged for the car to meet us at twilight; and it was a severe blow to learn that we had been misinformed at San Vigilio, and that no steamer of any sort would leave Riva that evening. The service indeed had been suppressed. Riva suddenly became the least habitable city on earth. It was shut in and encased and stifled by prodigious, fantastic hills; its principal hotel lacked soap and towels. It was too primitive to be even romantic. We might have hired a car to run us to Malcesina by the beautiful new road which joins Malcesina to Riva. But the beautiful new road had not been opened to traffic, because Verona was litigiously objecting to it. Verona divined in that thoroughfare a means by which travellers could escape the necessity of passing through Verona. It saw in the road a menace, and it was somehow fighting the road. Of course the road will win. History was repeating itself, and will continue to do so.

We might have driven right round the lake by the western route, but the distance was considerable, the route was dangerous and dizzy, and the night would be dark. The alternative was to hire a motor-launch for Malcesina. The harbour was empty of launches. Then we approached an old man of marine aspect, who said at once that he could and would supply us with a launch. Heaven! Escape from Riva! What, however, he really meant was that he could introduce us to somebody else who would supply us with a launch. Doubts! Apprehensions! San Vigilio seemed as far off as the other shore of the Atlantic. We thought mournfully of our car waiting for us at Malcesina, and waiting and waiting. Withal we did meet the second marine person, and he led us to another, hidden harbour, and introduced us to a third person who was the actual chauffeur of an actual launch, which we beheld there before us. Both the second and third persons were delightful youths—gay, smiling, laughing, leaping about.

But the price for the short trip was 200 lire. I demurred. They exhibited a printed tariff. I yielded. Not for 2,000 lire would I have picnicked in Riva that night. We embarked in the launch. Accommodation for accommodation, and distance for distance, the launch was dearer than the White Star Liner *Majestic*. No matter. Its name was *Veloce*. If speed is to be measured by splashing and noise, the *Veloce* was well named.

The lake was rough; the mountains were

darkening; we had a fresh head-wind. Each time half the lake fell into the launch the engineer looked round and grinned. (He himself was well protected.) The launch hugged the mighty cliffs for shelter. The lake broadened, and as it broadened its waves diminished. . . . We descried the castle or palazzo of Malcesina. We made the harbour. Forty minutes for twelve miles. Yes, the launch was worthy of its name. Nevertheless a tariff of five lire a minute did appear rather excessive.

We disembarked under the arcades of a hotel, whose porter reckoned us as guests for the night. Too good to be true that the car should be there awaiting us! But there it was. We drove home on a narrow road through village after village, all prematurely lit with cheapest electric light, and all restfully complacent in the feeling that the day's work was finished. Only a few small boats were fishing in the dusk—perhaps for fun, as the British husbandman tills his garden for fun after a hard day's tilling for his employer! We reached home. Riva seemed as far off as the other shore of the Atlantic.

Verona.

Juliet's home-town, I suppose some would call it. The phrase takes the edge off romance, and I designed it to do so, determined as I am somehow to vent my rage at being shown Juliet's house, a picturesque and untidy tenement, with balconies certainly too high for love, unless Juliet was a trapeze acrobat, accustomed to hanging head downwards by her toes.

This was not Juliet's house, for the sufficient reason that so far as authentic history knows, there never was any Juliet. It seems that Shakspeare took the story of Juliet from an Italian fiction, performing in the process his customary feat of making a silk purse out of a sow's ear; and that he chose Verona for her habitation because of its agreeably sounding name. There not having been any Juliet, there could not have been any Juliet's house. Hence to label a building as Juliet's house, and to draw the special attention of simple-minded tourists to it as such, was an act of unscrupulous fraud, which the city authorities ought to have firmly and publicly disowned. The thing is as barefaced a swindle as the alleged tomb of Agamemnon at Mycenæ.

Still, Verona is not a mean city. The first view of the vast old reddish Castle, with superb bridge to match, after you have passed through its triple enclosure of walls, is exceedingly impressive. Also the place is unspoilt. It exists now as it did exist. The hoof-marks of the globe-trotter are not upon it. The streets are narrow, with very few new monuments, and without vistas. (True, the traffic is directed by policemen with white batons, derived doubtless from London via Paris via Rome and Milan.) The main street is forbidden to all wheeled traffic. The famous Herb Market, where the original frescoes on the façades of the houses largely survive, makes a truly romantic spectacle. The one defect of the Herb Market is that the supply therein of the modern staff of life—I mean oranges—is both insufficient and inferior.

And the citizens are unspoilt children of the Renaissance, ingenuous, provincial, violent in face and mien, unpolluted yet by their brief contacts with the touristic horde! We were the only Anglo-Saxons to lunch at a purely Italian restaurant under the immense arcade of the old Bourse, where the waiters dashed to and fro in ordinary suits, arguing with one another, and being charmingly explanatory to their strange guests. We took coffee at the best hotel, and tea at the other best hotel; and both were pretty third-rate—according to touristic conventions. I liked that, especially as the tea was good and the coffee was good. You seldom or never get a first-rate hotel in an unspoilt town. First-rate hotels arise on the ruins of the primitive.

I was pleased, too, to see that no concerted effort had been made to utilize socially or touristically the fearsome river Adige, on which the city is situated. Not a terrace on its banks; not a café with a river view. Evidently the citizens regard the swift-flowing Adige merely as an impossible stream. They have imprisoned its turbulent water between granite walls, and then just left it to rafts and lumbermen. The principal open-air cafés are all in an inner square, where you can see girls walking and officers walking (Verona is the headquarters of an army-corps) and the inhabitants behaving naturally and self-unconsciously, as they do in Seville.

At the lunch hour there is an enormous rushing outbreak of bicycles, which though dangerous to limb are less so than automobiles and belong more to the historic past. Neither English nor French is spoken in the dark and cavernous shops. And as for the Post Office, which you reach by a splendid balustraded stone staircase in a splendid courtyard, it may be called Renaissance in its routine as in its architecture. The employees appear to say to each other: "Funny thing! Here's somebody wanting a money-order! What next, I wonder!" All which is delightful and touching—in homeopathic doses. The huge Roman gateway, the huge Roman amphitheatre—well, you can study them if you so desire. The inhabitants negligently tolerate them. And if you have been travelling a lot, you yourself are as sick of gateways and amphitheatres as of churches. I remember an old retired wealthy industrial Frenchman remarking: "Italy would be tolerable if it were not for its public monuments."

You leave Verona, after your first visit, with an impression of tremendous beating sunshine and of a general higgledy-piggledyness. The people have the phlegmatic indifference of Englishmen at home. They may just have heard of Shakspeare. If they have heard of the fabled Juliet they assuredly set her down as a wanton wench, imperfectly guarded, who deserved all she got from destiny.

Verona.

Second attempt. We visited the old reddish Castle, after a merciful fate had prevented us from seeing the museum of modern pictures. Its barbaric interiors well illustrate the life of the great days of Verona. Wonderful it must have been that any man survived the age of thirty. The endless picture-galleries present the usual appearance of all foreign galleries except the finest. Twilight everywhere. The pictures dirty, and under-framed and oddly labelled. Several Mantegnas of the second order. A Bellini ditto. A first-rate portrait by Titian. . . . And then you see a sculptured Virgin and Child, the Child with its finger in its mouth. It is labelled Donatello, and probably no one but Donatello could have done it. An enchanting work. A gem. The sight of it transforms the whole of Verona, makes Verona richly worth while. It completely changes one's attitude towards Verona as a centre of art. Unhappily it is crumbling to pieces, uncared for. But the citizens of Verona will bear this misfortune with an admirable tranquillity.

One church, San Zeno Maggiore, in a squalid suburb, had to be inspected, no matter how serious your surfeit of churches. It has the reputation of being the finest Romanesque church in North Italy, with the oldest bronze doors, and it possesses a celebrated and large Mantegna picture, the Madonna Enthroned. Impossible to keep out of this renowned church, which was begun in the eleventh century—and too conscientiously restored in the late nineteenth.

The Mantegna was worthy of its legend; the Roman and Egyptian pillars in the crypt, in which lie the immortal remains of San Zeno, fisherman and bishop, were exciting; the whole building had style. But the most interesting phenomenon in the great church was the guide, who captured my affections to the extent of ten lire. This extremely unusual guide had learnt no lesson by rote. He talked vivaciously in both Italian and French, and at once communicated to you his personal passionate enthusiasm for the marvels under his charge. He begged you to stay as long as you liked in any spot you liked. He knew when to be silent. He had none of the usual guides' air of having designed, built and decorated the fane with his own brain and hands. He talked as an artist might talk. Were all guides as the guide at the church of San Zeno, I would be ready to revisit every public monument in Italy. I have met his equal only once—when I chartered a taxi in Hanover Square and asked the driver to drive me to St. Bartholomew's. The driver's face immediately irradiated by the mere mention of that name. He drove me to St. Bartholomew's. He knew it by heart. And he knew all the other City churches by heart. He was a fervent amateur of City churches. And as I left him, with reluctance, in Hanover Square again, his smile indicated that in his opinion I had been conferring a favour on him by learning from him. The world is full of strange, lovable characters.

Once in the bar of a French seaside hotel the barman was mixing a Bacardi cocktail for me. Now I had previously seen this barman trying the violin of the leader of the orchestra which at night enlivened the bar. He played the violin with decidedly more emotional quality and with a more beautiful tone than the leader of the orchestra himself. " So you play the fiddle," I ventured, in English. " The fiddle ? " he questioned. " The violin," I corrected myself. " Ah yes, the violin ! " " You like it ? " I asked. " I love it," said he. " You are from Paris ? " I asked. " No, sir, I am Hungarian." He sighed. Discretion, a sense of decency, forbade me to ask him : " If you play the violin, and so well, why are you behind a bar mixing a Bacardi cocktail for me ? " A day or two later I learnt from somebody else that despite his passion for, and proficiency upon, the fiddle, the young man's real ambition was to be a dentist, and that he had already passed the preliminary examination at Geneva to this great end.

Turin, July.

Turin is a very large city, rectangular in form and spirit. It is so large that when in the centre of the city I asked the concierge of the hotel where were the most important and amusing cafés, he gave me detailed directions and then said: "It is twenty minutes' walk." So I did not visit the most important and amusing cafés of Turin. The city has the reputation of being full of public monuments. But the sole public monuments I saw were ambulatory ones—tram-cars, screeching on curves, thunderous and innumerable. I was informed that they cease to run for a few hours in the middle of the night, but I doubt it.

Superstition still dominantly survives. While I was eating a very late dinner, the waiter asked me the number of my rooms. Since I knew that the number of one of these was 12, I answered "Twelve and thirteen." "Twelve and fourteen," he corrected me. "There is no thirteen. Thirteen has been forgotten." He smiled. Quite possibly there is no No. 13 in any hotel, but I had never heard before of this delicate compliment to the sensitiveness of hotel-guests. Perhaps if there were a room No. 13 in a hotel, the proprietor might be able to let it only about once a month. Anyhow, we are living in the twentieth century, not the tenth, and in Europe, not in Central Africa. And a large proportion of the educated populace still believes that by psychic powers it can foretell the destination of a roulette-ball, and the number of pips on cards before they are turned up at *chemin de fer* or baccarat. And if you laugh at these persons they reply that you are not spiritual but earthly.

It is astonishing how hotels will presume on their prestige, especially in commercial towns. On inquiring for rooms at the biggest and costliest hotel I was shown two at 18*s*. apiece, plus something, plus something else. When I demurred at the price the exhibitor said he would accept 16*s*. apiece. This graciousness aroused in me an unchristian hostility. I refused the rooms at any price; their aspect had nothing to inspire lyrical praise. I was then shown two others, at 9*s*. apiece. Being somewhat exhausted, I weakly took them both. They had no running water, either hot or cold—the first bedrooms of the kind I had encountered during a month's incessant travel in France and Italy. On the other hand, they had vaulted ceilings richly decorated. While you are waiting for your hot water in the morning you have leisure to admire the ceilings.

Another instance of presuming upon prestige occurs to me. In a celebrated hotel in a celebrated town in the Burgundy country, I had an admirable dinner and the finest Beaune I ever tasted. The wine was served in enormous glasses, as a liqueur brandy is served in England. I unfortunately lost my self-control and drank more of it than became an ascetic. However, it was worth the uncomfortable sensations of the next morning. The wine-waiter was well pleased with my remarks; and the entire staff had an air of self-complacency. But the landlord was presuming upon his cuisine and his wine. My bedroom had not a single hook, nor a hanging wardrobe, nor a single drawer save a little one in the dressing-table. All the private bathrooms had already been taken, and there was not one public bathroom in the whole place. The interesting question is: when will the landlord be compelled by the march of time to cease presuming?

Italy.

There are a fair number of hogs (species: road) in Britain. But there are far more in France (subspecies: owner-driver). The French owner-driver shows a strong objection to being passed, and an equally strong predilection to pass. He lacks the social sense, the goodwill, the large benevolence of the English driver. I have been in two slight accidents due to the ruthless egotism of the French owner-driver. The second one arose out of a lack of co-ordination between the owner-driver and his female companion. The pretty thing gaily waved to us to pass. Thus encouraged, we passed; but the owner-driver curved his resentment at the affront in such a manner that there resulted a slight shock and the abrasion of one of our rear mudguards. What happened to the passed vehicle we shall never know, for we kept straight on, despite its hooting. We noted, however, that the machine was still moving. We had the intention of taking tea at Aix-en-Provence, a few miles ahead; but prudence caused us to postpone our tea. Though the French telephone is less efficient even than the English, if somewhat more so than the Italian, it still has possibilities—in the hands of a resentful owner-driver.

My first trifling accident, on the Riviera, had matured into such a pother and cacophony of languages English, German and French, even in the absence of the police, that the prospect of the intrusion of the French police into a motor-accident was too affrighting to be contemplated. Not to stop was a social misdemeanour. To stop would have been something in the nature of suicide. The greatest of all laws, that of self-preservation, swept us swiftly through Aix into a side road—which we followed, not by intention, but by error.

French road-signs are admirable in theory but lamentable in practice. On very many roads there is a stone at every hundred metres ; and the kilometre stones are nearly all marked with the names of the principal approaching towns and the distances thereof, nicely calculated to the nearest hundred metres. What purpose is served by the 100-metre stones I have never learnt. The purpose of the kilometre stones is apparent, but is rarely served. The stones are too low ; the lettering on them is not currently legible : and in the rich weedy season the entire stone is too often obscured by the luxuriance of nature. The blue signs at cross-roads and by-roads are loaded with topographical information, which however can only be studied by stopping your car dead and using a microscope or a telescope, or climbing up the post.

On the other hand, in much-frequented districts

gigantic and unnecessary information is imparted again and again, with a most exasperating iterance, and proprietary brands of automobile accessories compete enormously with one another in telling you what you have already been told one thousand times.

In Italy the road authorities are more logical and more laconic. They indicate direction once, in largest letters, and so long as the main track is plain to an average intelligence they don't indicate direction a second time. (And there are few milestones, and still fewer readable milestones.) You may drive ten miles without a hint of direction, but when the moment arrives for guidance, the guidance is always there—in large letters. This is a nerve-soothing system, and fosters the excellent philosophy of fatalism.

Regarded as a whole, no road-surfaces are equal to the English. But some French road-surfaces are almost as good as the English, and as a rule French roads are better planned. Many of them were re-designed after the Revolution, when authority *was* authority, and if a road was better straight, it was put straight, and landowners had to like it or lump it. Thirty years ago when I used to bicycle about France I found the straight stretches of road endlessly monotonous ; but they are not a bit monotonous in a car. Italian roads are fairly well planned, and often have fairly good surfaces. But the absence of tar, and the superabundance of dust—particularly the superabundance on Sundays—make for acute discomfort. The *autostrada* or speedway between Milan and the lake districts has a surface as good as the best English; and it is straight and level, and the price of it is roughly a penny a mile, and you can travel on it as fast as your engine and your rashness permit. But there are houses and infants here and there on its margin, and the scenic effects are nil. The *autostrada* is a great time-saver. Its kilometre-stones flash by like telegraph poles.

There is a fair amount of countryside defacement in England. There is far more in the fashionable regions of France. The horridest rows of County Council houses are masterpieces of classical architecture compared to the domestic edifices of Deauville, Le Touquet, and the Riviera. The main Riviera coast-road (surely the most dangerous in Europe, with its double tram-lines, its ever-curving narrowness, and its motor-buses merrily overtaking Rolls-Royces at blind corners) from St. Raphael to Mentone is now an uninterrupted series of architectural extravagances equalled only in the Los Angeles region where, as I have been told, Spanish palaces have been raised by sticking stucco on to chicken-wire for the accommodation of waitresses who, having become film-stars in a night, arrive gloriously to inhabit them with vast retinues of fathers, mothers, cousins, chauffeurs, butlers and maids.

No region, perhaps, was ever more miraculously endowed with natural charms than the Riviera, and certainly none was ever so disfigured by the vulgarity of man. Every time I visit the Riviera I take oath that that time shall be the last. But it never is. If you have your own house there, in a large withdrawn garden, and do not leave the domain save occasionally at night to enrich the proprietors of gambling-dens, the Riviera is habitable. Not otherwise.

When we were descending into the port of Nice with its sober traditional architecture, a friend said: "This is like going back to France." Then the relief of entering the architectural sobriety of Italy! On the Riviera there are ten times more roadside advertisements to the kilometre than in England to the mile. The spectacle is absolutely desolating, and if its end is to fill lunatic asylums, then the end is no doubt achieved. How can it be ameliorated? It cannot be ameliorated, save by public opinion, and there is no public opinion. In Italy road advertisements are fewer and less hideous, and the owner-driver is more courteous. But advertisements are not unknown. Every town and village is announced as "Shell" or "Lamp", or both. Which reminds one of the age when all Underground stations of London were called "Partington".

Crossing the Alps, July 20th.

Never having accomplished this feat before, and thinking thoughts of Hannibal and Napoleon, I approached the region, in the direction opposite to Napoleon's, not without secret excitement. It was exceedingly hot in Turin. It was exceedingly hot on all the long straight roads leading to those notorious mountains, and as soon as we began to climb them, it grew still hotter. The radiator of the car reached boiling-point again and again, and had to be refreshed by icy water transferred into it from roadside brooklets by means of my hat. Otherwise the upward journey was simplicity; notable only for the numbers of cataracts and torrents, and of small companies of cows in charge of children. The road was conceived by Napoleon's engineers in a broad-minded spirit; the epithet "generous" has been well applied to its hairpin bends. The gradients are severe, and consistently severe, for miles, but any modern car would take them with perfect nonchalance; and the surfaces are excellent.

I imagined that something worse would come; I was apprehensive wrongly. After a brief stretch of steady ascent, we perceived that the Alpine summits were strangely close to us; indeed, so close, in the thin clear air, that it seemed as if by putting out an arm one might touch the great patches of snow on them. The sunny heat was still intense, but there lurked under that caressing warmth some kind of a chill menace. Also here and there by the roadside ran short tunnels—avalanche-galleries for the protection of wayfarers, and at intervals were refuge-huts established by mountaineering societies for the needs of adventurous persons caught by the night or the storm.

All of which struck one as odd, and perhaps disconcerting. No danger whatever, in the July sun; and yet deadly danger was waiting, hidden amid those peaks, ready to pounce down upon travellers in other seasons and weather. One had an irrational sensation of insecurity.

Then an official warning stopped us: warning not of death but of the Italian Customs, alert to examine the possessions with which we were attempting to leave the country. Difficult to believe that we were already practically at the top of the pass, but we were, and 6,000 feet in the skies. After all, what is 6,000 feet? The whole city of Johannesburg is as high, and no one thinks twice about it. The chief of the customs came out to me in the car, and spoke in English, and I shook hands, being made instantly aware that he had ideas on English literature which he was anxious to express to a writer. He expressed these with unsurpassable courtesy, and I expressed one or two of mine with British clumsiness, and we parted. The too brief interview was charming and highly ceremonious. I would have talked on; but no, he could not dream of delaying us, and with bows he re-entered his office which was also his home. I have often thought of him since, eagerly studying all manner of British authors, and those authors not in the least aware that his Latin eye was upon them, peering critically down from the Alpine eyrie.

In a few minutes we came to a large, low building

that resembled a fortress. We went in at one gate and out at another, and though soldiers were wandering around none accosted us. We assumed that we were in France. And we passed also a large tarn: not a boat on it, not an angler on its banks; and for good reason. The mere look of its pale blue water appreciably lowered the thermometer which every one carries in his heart. In winter the lake must be frozen very thick. It was the most deserted, frightening, useless thing you ever saw.

Somewhere either before or after this—I forget; the frontier arrangements are confusing and incomprehensible—we had to stop again and go into a wretched ramshackle cottage, where three Italian officials gazed at us sternly, inspected passports, and asked for detailed information on topics not referred to in the illustrated passports. Our parents, long since dead, were exhumed for their curiosity. And one of the officials, seated at a rickety small table amid the squalor of the bare room, wrote down our history and the history of our parents, with a foul pen; and as he knew little French and no English, he certainly got the history all wrong; and anyhow nobody would ever read his historical work. This original research occupied a long, long time.

We foresaw that if the French officials were equally laborious we should arrive nowhere for lunch. But there were no similar French officials, and I assume that the Italians, by some fraternal concord, were acting for two great nations and not one only. The Locarno spirit! We ran swiftly over perfect surfaces down easy descents for twenty minutes into indubitable heat, and were congratulating ourselves on a complete emergence from the dark moral oppression of officialdom, when at a little town, Lanslebourg, an official automatically applied our brakes for us by stepping in front of the car. The French customs, miles off any frontier! Still later we reached Modane, and more heat, and saw railways and the disappearance of a railway into the flank of a mountain. Burrowing through the Alps presented itself as a much more formidable business than climbing over them. Crossing the Alps may have daunted Hannibal and Napoleon; but to us it was nothing. We had crossed them, and interviewed literary critics and official historians and soothed the carping mood of other officials, all in a couple of hours.

I should be interested to learn why the Mont Cenis tunnel is so called. It certainly does not run under Mont Cenis, nor anywhere near Mont Cenis.

Avignon, Genoa, Milan, Bergamo.

The contrast between old Bergamo on the fortified hill, and new Bergamo on the edge of the vast plain, is really rather dramatic. New Bergamo is just new and naught else, with hotels containing "the last comfort", ostentatious public buildings, broad tree-lined avenues, and policemen at cross-roads.

Old Bergamo is completely old. Nothing in it is modern, except bits of sanitation here and there. It is the essence of the natural-picturesque. Its fanes look miraculous in photographs, and also in the stone-flesh. Its piazza has two public libraries, but whether anyone ever goes into them I know not. I didn't. It has a fountain, little café-restaurants, a little hotel, some shops, a barber, some quidnuncs, and a few children who spend half an hour in staring at you and then rush off like red Indians.

We were saying how delightfully one could live there for ever and ever. But one couldn't live there in delight. Because one's mere permanent presence would be unnatural, and would ruin the intactness of the place. A grand hotel stuck like a toy on the huge flank of an alp is unnatural enough. The French Riviera is unnatural enough, though its unnaturalness is fast becoming the natural. But British inhabitants in old Bergamo, with their craze for modernizing the interiors of houses, their morbid insistence on comfort, and their repudiation of the environments in which they plant themselves, would be an offence.

We ate lunch and drank good coffee on the piazza, viewed the monuments, flaunted a red guide-book, purchased postcards, and descended the steep, narrow cobbled streets to our proper refuge, new Bergamo.

Travel is full of startling contrasts: one of the most startling is that between the richly decorated, idle luxury of the large Italian liners in the port of Genoa, and the slums and never-ceasing toil which surround the port. Not for a single hour in the twenty-four is the port quiet. Another contrast is that between the magnificent suburbs of Milan and the squalid suburbs of Genoa. The splendid straight smooth roads radiating from Milan easily surpass those of London or Paris. The periphery of Genoa is terrible. Why? Genoa is a very important place, and is indeed called, justly, the "city of palaces". One hears that the relative rise of Milan is due to the influence of a certain Alpine tunnel. But there must be a more complex explanation than that.

And yet another contrast. We sat in one of dozens of large, too-musical cafés in the centre of Milan. All full; and the pavements full, and the tram-cars. Tremendous babbling crowds, of which an extraordinarily large proportion consisted of smartly dressed young men. An enigma of the life of big cities is the multitudes of young men therein who apparently are free to enjoy themselves at four o'clock in the afternoon. (But you do not see them in London, nor in Manchester nor Glasgow.) From the cafés of Milan we walked a few yards into the cathedral. Hardly a soul on the measureless desert floor of the unornamented house of God. And after the sunshine glitter of the streets and squares, the sacred interior seemed to be in that final stage of dusk which immediately precedes black night. Darkness and utter silence. In a distant corner, an altar was lit with candles that gave an effect as of fireworks in the heavy gloom. The sudden change made you think, almost overwhelmed you.

But it was at Avignon that I encountered the greatest contrast. I do not care much for Avignon. It is over-visited. The Palace of the Popes is not worthy of its reputation, and the inhabitants are not sympathetic. They must suffer dreadfully from the mistral. I once suffered myself from the mistral in Avignon, and I shall not forget it. The mistral is a wind to destroy nobility of character and dry up the milk of human kindness.

We stayed in one of two largest touristic hotels. The evening heat was very oppressive. The hotel was full of assorted British and American accents. Not a " foreigner " in the building, except the mistral-cursed staff! I often wish that all Anglo-Saxons (except me and my companions) would have the decency to stay at home and leave the continent of Europe uncorrupted by their presence and voices. The crowded dining-room of that infernal hotel was more Anglo-Saxon than London, and much more Anglo-Saxon than New York.

We fled from it, and in about three minutes were in the square of the populace. Some Anglo-Saxons would have the effrontery to call it the native quarter. The large square was a most soothing spectacle. All cafés, theatres, cinemas. The mundane name of Esther Ralston written in electricity on a façade, and under it two black-robed Italian young women taking their ease in wicker-chairs and gossiping with a man in evening dress who bent laughingly over them. The young women must have been ticket-girls of the cinema, and the man may have been the director. Groups of natives drinking and chattering on the *terrasses* of the cafés. Crowds of natives sitting or sprawling on the hot pavements ; and many children among them, including a baby of one and a half or so, who propelled himself to and fro on his basis with considerable skill. The hour was ten-thirty. French children seem never to go to bed. Yet they are beloved, petted, cared for, and look well. Evidently the British method of rearing is not the sole satisfactory method.

We had a glimpse of a corner of the Palace of the Popes, just beyond an open corner of the square. The architecture, scarcely visible in the night, rose majestic above house-roofs : as mighty and magical as a stage-set by Gordon Craig. The vision drew us up to another great square, dark and quite empty, and then into the public gardens on the far side of the Palace. We heard the ringing of a bell. A guardian arrived and turned us out of the gardens jocularly, and locked the heavy gates behind us. Eleven o'clock had struck.

Back in the first square, where everything, infants included, was proceeding just as before. We sat down on the terrace of a café, and ordered drinks, polluting the nativeness of the place by our touristic aspect and deportment. When we reached the hotel, it was dark, and the gate was locked, and we had to ring more than once for readmittance into Anglo-Saxondom. Anglo-Saxondom was in bed and no doubt as fast asleep as the heat would allow. Our sightseeing fellow-guests had missed the most interesting sight in Avignon.

Albi.

Something about the name 'Albi' which inspires feelings of romance! For many years I had had (what a woman would call) an intuition that the cathedral at Albi would be marvellous. It was so marvellous that last evening on arrival we visited it before even going to a hotel to look for rooms. There may be other cathedrals in a similar style; I have not seen one. Ruskin ought to have described this cathedral. For the present unfashionableness of Ruskin makes no difference to the fact that he was a really great describer. Of course he would have moralized, and shown how every piece of the fane illustrated the character and ideals of the people and the nature of their history; but he would have described it. (Perhaps he did. I sold my elephantine edition of Ruskin before the war.) At an early hour this morning I revisited the cathedral. Its red brick is eternally fixed in my mind.

Montluçon, July.

I was looking for rooms in a hotel near the railway. As the landlady and I went along a corridor upstairs, I said I hoped the noise of trains could not be heard in the bedrooms. "One hears nothing," she answered positively. She opened the door of a room, and a tremendous engine-shriek met us, seeming to drive us both back from the threshold. She shut the door, and tried another one, and we were met instantly by another tremendous engine-shriek. She burst out laughing. I laughed too. If she had not proved her sense of humour I might have walked straight out of the hotel. But her sense of humour kept a customer with a sense of humour. I got quiet rooms at the back.

Montluçon is rather an ordinary provincial town, which probably few stay at except business men and the benighted. But it has a somewhat elaborate system of traffic-directing policemen. Provincial towns, both in France and Italy, seem to glory in the profuse provision of agents of order; also in one-way streets. Municipal snobbishness: that is what it is. Traffic-directing policemen and one-way streets are clear proof of the existence of a large volume of traffic. Moreover, small towns must imitate large towns and the metropolis. At Montluçon we asked a grim, sardonic policeman the route for Le Puy. He gave us an answer than which nothing could have been more wrong. After two miles we had to return into the town. One of us said that the incorrect policeman had acted from mere provincial maliciousness, hatred of the tourist, etc. I would not agree to this explanation. Obviously the policeman could not admit ignorance. He just made a guess and the guess happened to be a bad one. We asked another policeman. He said: "Follow the high street *and ask again of the fourth policeman you see.*" We did so and were saved.

Poitiers, Saumur, Angoulême.

At a restaurant entitled the Chapon Fin, I had one of the most wondrous meals of my existence as an eater. And not too dear. Poitiers is only a small town, and yet the largeish restaurant was full of lunchers all doing themselves exceedingly well; and few tourists among them. Of the few tourists the most astonishing were two English or Scottish sisters. They ate at length, and never spoke to one another. One had a book and the other a newspaper. They ate truly distinguished food, dish after dish—and they read. Nevertheless, if you tell a woman that women don't understand food, she will deny the criticism with resentment.

Saumur is an enchanting town, full of fine domestic architecture. Angers is an enchanting town, full of fine domestic architecture. So is Angoulême. But all three are very provincial. As we were driving in the rain through Angoulême, a lady in my party suddenly exclaimed: "Oh! There's a hat!" The car was violently stopped. She dashed into the shop and emerged with a 300-franc fashionable green hat, snatched from the provinciality of Angoulême. What an eye she had for a hat! We drove on.

Blois.

Admiring the beautiful façades of the château again, after long years, we decided that we might as well visit the interior, which I had never seen. "Only five minutes," said the guardian of the door, after we had paid the fee. The hour of midday closing was about to strike. Five minutes was just enough. One can get a first general idea of any château interior in five minutes, and of any picture-gallery in half an hour. Details should wait. It is the large impression which ought to come first.

We saw a group of people in front of us, and the herd-instinct drew us towards them. In a moment we were caught, held, by the eye of the official guide. He locked a door behind us, and unlocked a door in front of us; and throughout the tour he kept locking doors behind and unlocking doors in front. We could not linger. We were ambulating prisoners, and there was no getting away from the ordeal. The ordeal was terrible; such ordeals always are. Official guides confine their explanatory remarks to the statement of centuries—this was fifteenth, that was twelfth and so on—and to childish details concerning secret closets and staircases and the thicknesses of walls, and to differentiation between what is original and what is "restored". They rarely say anything of interest. They are decent fellows, but self-complacent. Most of them have gradually been victimized by the extraordinary delusion that they themselves are somehow creatively responsible for the wonders which they exhibit.

The highly uninhabitable interiors of Blois have lost all their furniture, and in both architecture and decoration they are merely barbaric. It is marvellous that architects so brilliant in façades should have been so dull, clumsy, and inept in planning and decoration. The thought of the darkness and discomfort of those kingly times depresses, and the depression is rendered acute by the absence of interior beauty, by the solemn, silly ritual of the guide, by the desire for freedom, and by the apparent stupidity of your fellow-tourists. They really do look stupid. But for all you know you yourself may look stupid, as you listen to the interminable, echoing rigmarole and wander on and on with the meekness of sheep, upstairs, downstairs, through guard-rooms, audience-chambers, bedrooms, boudoirs, dungeons, all cold and damp in the intense dry summer heat. It is my theory that the compulsion to listen to what is stupid must induce stupidity in the listener. The one strong argument against this theory is that the guide spoke in French and obviously the majority of his restless defeated flock had no notion what he was saying. Nor had they any eye for the occasional glimpses of lovely exteriors—the carving of the rows of gargoyles, for instance—as to which the guide uttered no word.

At length, when you had begun to long ardently for an earthquake or the end of the world, the guide said: "And now ladies and gentlemen, the tour is terminated." A lock creaks. Glorious prospect of liberty! You have, morally and in fact, to tip the guide for half an hour of mental and physical torture; but not even this horrid necessity can impair your joy and your relief. You had walked like sheep. Now you skip like lambs. You dash down a dangerous stone staircase. You are enfranchised. You are in the sun, in the courtyard, in the street. The adjacent too-touristic hotel where you will lunch bears a convincing resemblance to Paradise.

Fontainebleau, July.

Yesterday I visited Fontainebleau. Another of those châteaux! All French châteaux are not on the Loire, and this one is not. It is very nearly on the Seine, which is beautiful here, especially a little higher up, at its confluence with the Loing. The French have not yet quite learned how to exploit either of these rivers for pleasure. What they chiefly do is to sit on them in a moored punt and fish with perfect futility for hours. The barge traffic, however, is interesting.

Thanks to the facts that Napoleon liked Fontainebleau more than any of his other residences, that he imprisoned the Pope there for nearly two years, and that he abdicated there, under stress, Fontainebleau is probably the most popular château in France, just as its superb forest is the largest. Every ten minutes throughout every day of the week, a party is conducted by a guide through the showy portions of the château; and as the tour takes about forty interminable minutes there must be three or four parties *en route* at once, each dominated by a guide who says the same things in the same words eight or ten times daily. These guides surpass actors in that they seem never to "dry up", and never to "fluff". True, the runs of their sedate performances must easily beat all theatrical records.

The mischief with their work is that they are evidently bound more strictly than other château guides to a time-schedule, a schedule which leaves no margin for leisure or caprice. You are moved inexorably on, for the parties behind you are hurrying near, and the concourse of visitors in the waiting-room is always growing, and growing impatient. The guide is succinct—he has to be—except when he indulges you in childish details, such as the exact number of different woods used in a complex parquet floor.

The rooms, as usual in châteaux, are badly planned, often dark and often inconvenient. The darkest and the smallest and the most inconvenient are those of queens; the most spacious are those of royal mistresses. Characteristic of masculine and feminine human nature! As surely as you enter a noble interior, so surely will you see the proud initials of Diane de Poictiers innumerably repeated in gold on the walls. Nearly all the rooms are cruelly overcrowded with furniture. Much of the original furniture is stored away in hidden depositories of the château, whose appearance would be improved if still a great deal more furniture were stored away. Though there are lovely things here and there, the general effect is one of an overpowering sumptuous vulgarity.

Last year a French journalist, novelist, and wit, Pierre Mille, put the silly-season question: What would have happened if on the 15th August 1769, Letitia Romalino, wife of father Bonaparte, had given birth to a girl instead of a boy? It may be said with certainty that one thing which would have happened would have been a marked decrease in the dreadful imperial ugliness of the château of Fontainebleau. Two of the most appallingly rich chambers in the world are Napoleon's throne-room and his bedroom—with the cradle of the King of Rome, whose misfortunes began with the sinister formidableness of the poor little dear's Empire cot.

I have a considerable affection for the Empire style, of which I bought a houseful when it could be bought for half nothing. But the excesses of the style are terrible, and Napoleon preferred the wildest ornate excesses. I had not beheld the interiors of Fontainebleau for twenty years, when it was far less visited than to-day. I enjoyed it less than ever. My estimate of Napoleon's culture—if he had any—was less than ever. My sympathy for the unhappy members of his court—read the devastating memoirs of the gradually disillusioned Madame de Rémusat—was keener than ever.

As for the guided tour, after skating, sliding, and slipping over a mile or so of polished floors you emerge with a mind confused, congested, and as utterly worn out as your legs. You have glimpsed everything and seen nothing—save the initials of Diane and the bust of Francis the First and the Pope's bed. You simply don't know where you are in history. Your crowded memory is a jumble of styles. Every room has at least two styles, and some have four. What is the necessity of a conducted tour? You are not conducted through Versailles, nor through the National Gallery, whose treasures are quite as valuable, and quite as susceptible to the depredations of the ungodly, as those of Fontainebleau. If a guardian were stationed in every room, or one for every two rooms, the need for conducted tours would cease to be; the visitor could see properly what attracted him and ignore what offended him; and the extra cost might well be met by increasing the price of admittance, which at present is a trifle less than fourpence a head. I should like to offer this simple suggestion to the most bureaucratic of governments, but probably I shall not collect the courage to do so.

From the château you cross the road to have tea—and never was tea more bitterly paid for in advance by hard labour—in a hotel which is fuller of Empire style even than the château itself. The last straw in an Imperial afternoon! . . . The experience of Blois repeated and raised to the 20th power.

Dieppe.

Caen may not have architectural monuments as fine as those of Rouen; but as a city it is the finer city—more complete, more homogeneous. There are terrible monstrosities in Rouen. A comparatively small city, Caen, no bigger than Hanley, the largest of the Five Towns. But it has all the air of a capital. And all the noise. And all the traffic and the police-controllers thereof. Also all the grades of hotels. I went into the swagger hotel, on the faces of the entire staff of which was written the word 'rapacity'; I could even see the word embroidered on the too-natty and too-trifling aprons of the chambermaids. When my baggage was unloaded I had it loaded again and departed amid unfriendliness, for I had jibbed at the prices asked. I went to another hotel. It was full. Then to a third, where I obtained accommodation as good as had been offered to me at the first, at less than one-third of the price. And the dinner was excellent. And it was just as easy to visit the expiatory churches of William the Conqueror and his Matilda from the cheap hotel as from the dear one. Curious how each of these churches illustrates the character of its pious, sinning founder. . . .

From Antibes to Dieppe (now a thousand times more English than Caen ever was) I motored 1,185 miles, averaging only 170 miles a day. And yet I seem to have lived in the car. On a tour one ought to motor only every other day. But one cannot. One is forced on and on by the distressing, irrational desire to arrive at a certain destination at a certain time.

Brittany.

Yesterday I was lunching in a strange hotel where I had never been before and to which I shall never return, seeing faces and hearing voices of mysterious human beings called waiters who knew me, and whom I knew, for an hour only. And while I was sitting in the lounge drinking coffee, and looking round at the expanse of the golf-links, whose existence is the sole cause of the existence of the hotel, I saw a procession moving slowly towards a windowed shed in a corner of the course. This procession consisted first of a policeman, second of a stretcher borne by two men, and third of a group of followers. Upon the stretcher I could make out nothing but a rug and a pair of heavy boots at the fore-end of the rug. The stretcher disappeared into the shed, at whose door the policeman stood on guard. The group of followers remained outside, staring. The number of starers increased. Then members of the hotel-staff went out into the hotel-garden and stared. And I stared. Then a waiter came up to me and without being asked for information said: "It is an Englishman, military. He was on the cliffs near here. He fell into the sea. They have just removed the corpse."

"An accident?" I suggested. "Oh, yes, sir.

An accident, without doubt. The cliffs are dangerous." When I left I passed close by the shed. The starers had disappeared, but golfers were golfing on distant parts of the course. Curiosity had exhausted itself. The door of the shed was open and the policeman stood in the doorway.

To-day I was informed positively that the affair was not an accident, but suicide. The English military officer was known on the coast for his peculiarities; a letter had been found . . . relatives were arriving from England. Yes, it was all very sad, very regrettable. A charming man, he was, though peculiar and of solitary habits. Etc. etc. What stuck in my mind was the pair of heavy boots protruding insensate, callous, ugly, from the fore-end of the rug . . . Mysterious existence of these eccentric self-exiled English! But not more mysterious than the existence of anybody else.

Brittany, July.

At 4.30, though the dawn has furtively begun, the lighthouse about five miles off is still flashing its red sector over the deep dusk; two flashes, a pause, two flashes, a pause, and so on without end. And the other lighthouse, about ten miles off, is flashing its white light. (Men on those lighthouses probably dozing!) The character of the sky cannot be determined in the gloom, except as far as it is indicated by a total absence of stars. I ignite the first and finest cigarette of the long day. A statesman of my acquaintance once told me that absolutely his first act on getting out of bed is to start a cigar—and his cigars are of the largest and most powerful calibre. After breakfast, for me, certainly a cigar. But before breakfast a cigarette suffices. The prime quality of statesmen is neither brains nor a too sensitive integrity, but rather a formidable tireless physique, joined to a gift of sleep which no professional anxieties can disturb.

I lean over the balcony, seeing naught save the warning lights, and reflect upon the reactions between my fellow-creatures and myself. Or rather, I do not reflect, for at that hour my mind is incapable of sustained logical thought. I muse, vaguely, meanderingly, reaching however the clear conclusion that the desired goal of moral perfection is still somewhat distant, and that I am often maladroit in my social relations and lacking perseverance in the pursuit of righteousness. In short, that there are better men on the revolving ball.

At 6.15 all is light, and the lighthouses—night-watchmen—have retired to rest and total invisibility. Every detail of the golf-course, where nature has been defaced in order to make gratuitous difficulties for the golfer, is plain to the sight. With my glasses I can even decipher the numbers on the white boxes of the teeing-off platforms. The restless sky is dark and terrible with steely menaces. The little flags on the putting-greens are fluttering desperately in a south-west wind. Not a soul. Not a sound.

Then a fox-terrier comes ambling along on some secret and no doubt sinister enterprise of his own, and disappears. He is the first soul. Then a youth on a bicycle free-wheeling down the gritty, bituminous road in front of the hotel. He jumps off at the door of a small building of which part is employed for the profitable business of the golf-professional. I look away; I look back; youth and bicycle have vanished. Then in the distance of the links I see two workmen separately approaching. Their paths are unnecessarily devious over the turf. I expect the paths to converge and the men to meet and talk, near the hotel. But no! The paths do almost converge but curve apart again, and the men give not a sign of being brothers in the same worldly boat. Perhaps they are preoccupied with household afflictions and the harshness of the coming toil. One continues up the hill rising to the north-west; the other enters the same building as the youth.

Work is beginning somewhere. A tiny fishing-boat stands out to sea, heeling over at moments to a squall. Work there! Then an aproned man appears underneath me from the portals of the hotel, shouldering a carpet which he lays out flat on the grass and sweeps vivaciously with a besom. He folds up the carpet in the same old folds, and carries it within again, trailing his besom. I look down on the balconies below me. Most of them are encumbered with highly-coloured bathing-costumes and white towels. The sun blindingly appears for a single instant between black clouds, and retires for the remainder of the day.

This strange blow of the sun reminds me that the hour is ripe for the eating of fruit. I eat fruit, slowly, lusciously, and either gaze at a distant castled island and other smaller islands silhouetted blacker against the black sky, or I read yesterday's newspaper. The tidings printed in the newspaper can only interest either the fool, the searcher after pleasure, or the student of mankind. I know that I am all three. Page after page of really the oddest tidings collected from the various "resorts" of Europe. I read of the arrival and departure of Count This and Prince That, whose wives are invariably American, of industrious committees organizing pleasure, of gaiety, laughter, and going to bed at dawn, of paper costumes so ingeniously contrived as to imitate silk, of competitions of beauty and elegance and even distinction, of small donkeys being hired and introduced into ballrooms to the end of intensifying the picturesqueness thereof, and of the thrilling expected arrival of a Prince greater than all previously mentioned princes.

Beyond a summary of international news and minute details of the fluctuations of every stock and share on every important bourse on earth, and a few photographs of celebrities and of young women more unclad than clad, there is nothing in the newspaper but those oddest tidings, which are assuredly intended to convey the impression that God's in his heaven and all's right with the smiling, prosperous, delightful world. I read and read, fascinated, mesmerised, enchanted. I can miss nothing. It is a sort of necromancy put upon me. The fruit and another cigarette are finished long since.

A man perched amid the whirr of a motor-mower is now operating on the putting-greens. Then a smart motor-car approaches. " *Voilà papa!* " says a voice. A lady in a peignoir, with a small child on her lap, is sitting out on a balcony near to mine. The car comes to a stop in front of the hotel, and a smart man, for whom the chauffeur has respectfully opened the door, steps down. " Say good morning to papa," says the voice. " Good morning, papa," squeals the small child. The smart man looks up, as in a daze. He descries his wife and child. " Good morning, little one," he answers in a tender tone, and walks into the hotel, followed by the chauffeur with his bags. And the lady and child disappear from the balcony. Where has that man come from, to arrive so smart and blithe at 7.15 a.m.? Heaven knows. Later, while I am still wondering, I hear a knock at my door. I leave the balcony. The chambermaid, black and white, all smiling and fresh, and bearing a tray, enters the bedroom. " Behold your complete tea, monsieur!" says she. " Thank you, madame," say I. She is a mere ten minutes late. I smell the unique, the revivifying odour of tea. I pour out the tea. I fall on the rolls and butter. The day has at last really begun and the world is fully alive once more.

Dinard, August.

We are perhaps too often unjust in our moral judgments. When I suggested to the manager of a hotel that 8 francs was a bit high for a simple bath, what I really and indignantly meant was that the charge was an outrageous ramp. The manager very mildly explained to me that he himself paid 5 francs to the waterworks for a cubic metre of cold water, and that the average bath meant half a cubic metre of water. He then had to bear the cost of heating the water, pay interest on the cost of the bathroom installation, pay wages for cleaning the bathroom, pay for the laundering of the towels, and provide for his overhead charges. I recalled a statement of the most efficient hotel-manager I ever met to the effect that his overhead charges came to 52 per cent. of his turnover. Thus, if he bought a bottle of wine for 4*s.* 10*d.*, he must sell it at about 10*s.* to save himself from loss. . . . I admitted that perhaps 8 francs for a bath was not such an outrageous ramp after all.

London, August.

Yesterday evening I took the night-boat to England. Lives of meek men all remind us we simply don't know what meekness is. " My " steward offered me his excuses for having recognized me from my photographs. Can this be surpassed? Also, I am astonished that anybody can recognize me from my photographs. But people do. Often I see them leaning towards each other and I hear them say, in a whisper audible a hundred yards off: "There's Arnold Bennett." Which I think is not quite the best manners on their part. Late at night I had a heart-to-heart talk with the steward. Like most stewards he was a pale, thin, quiet man.

"Do you get any sleep at nights?" I asked him. He answered: "Well, sir, it depends on the season and the passengers. Sometimes *we* [never 'I'] get a bit of sleep. But of course it's night work. If I'm lucky I'm off duty at nine in the morning, and when I'm in England I go home to my wife and family, and sleep."

"But holidays?" I asked. He answered: "Well, sir, *they* give us a week's holiday in the year." "But Sundays?" "We work Sundays as well as weekdays." I said: "But do you mean to say you work every night in the year except seven?" He answered: "Yes, sir. It's all the same, week in, week out. Of course it's monotonous, but that's the conditions and we accept them." I said: "But you ought to have a night off a week and a fortnight's holiday." He smiled: "Well, sir. That's what it is." "Well," I said, "it ought to be altered." He said, indifferently, fatalistically: "Yes, sir."

He was a very agreeable and soothing man, but, though yet young, he had sunk into a groove extremely deep, and had apparently lost all divine discontent. He lived the life of a monk—and of a shuttle, to and fro, to and fro. Saw many dawns, but rarely the midday sun. Now and then an afternoon movie, never a theatre. He was cut off from mankind, except for glimpses of his family and the nocturnal society of his fellow-stewards. To me his existence presented itself as terrible, and the worst of the affair was that to him his existence was perfectly ordinary.

I retired to my cabin and slept. At five-thirty the next morning he came all bright into the cabin with my cleaned shoes and some fruit—and apologies for disturbing me! I bade him good-bye on the quay. Probably I shall never see him again. And in less than fifteen hours I have forgotten his features.

At Sea, August.

I am in an ocean steamer, a liner. I said to my steward: "How early can I have my breakfast?" He said: "Any time *you* like, sir." "Yes," said I. "That's all very well, but when do you start work? I don't want to inconvenience you." He said: "I am called at five and on duty at half-past five, sir." "Every morning?" "Yes, sir."

I began to be interested. I asked: "But when do you go to bed?" He said: "Eleven o'clock every other night, and ten o'clock every other night. Sometimes I manage to get off at nine-thirty." I said: "But do you get enough sleep?" He evaded the question. Here was a seventeen- or eighteen-hour day, seven days a week. I did not care to question him further. I knew that he was on duty all day, because whenever I rang the bell he answered it. I did not inquire about his holidays. Like the steward on the cross-channel steamer, he was pale and thin and meek! but a trifle less meek, and certainly more sturdy, morally. This is the day on which "everyone" in England either has gone to the north or is going. There seems to be some truth in the adage that it takes all sorts to make a world.

At Sea, August.

The same steward further informed me that if the interval between two voyages did not exceed four days, the Line kept him on. But if it exceeded four days, they got rid of him. He was anticipating an idle interval of three months soon. "What do you do then?" I asked. He said: "*We* go on the dole, sir. The Line sometimes loses good servants, sir, that way. We ought to get a small retaining fee." At the price of a small retaining fee, he seemed quite willing to be thrown out of work whenever it suited the Line.

I found on the lunch-table a printed card from the captain inviting me to attend boat-drill at 5 p.m., with life-belt. I at once decided that I would not attend boat-drill, not because I was not very curious to see it, but because I knew I should feel uncomfortably self-conscious, foolish, clad in a life-belt. So constituted is the nature of man that I preferred the risk of drowning at sea through ignorance to the constraint of appearing before my fellow-creatures in a cumbersome life-belt.

However, when the time came, I sneaked up to the observation-deck, and looked down upon the boat-drill. Sailors and stewards, all belted, stood in lines. Perhaps a dozen boats were swung out. They were swung out very slowly and awkwardly. Total absence of smartness in this drill. The ingenious machinery was stiff and it creaked a lot.

Not a boat was actually lowered even a foot. The distance to the water was some thirty feet. The sea was smooth, the ship steady. I wondered, if in such ideal conditions the drill was so slack and slow, what would happen in a storm, in a collision, in a semi-panic. I tried to imagine a picture of the lowering of the boats into a rough sea, or into any sea.

Up to now I had not seen a single passenger at the boat-drill. I was told that passengers were assembled on the deck below the boat-deck. And they were. I descended and beheld them, crowds of them, all self-conscious and falsely jolly in life-belts. One or two petty officers were about, and occasionally they adjusted a life-belt for a child or an attractive girl. Nothing more. Whether the boats, on their way into the ocean, were halted at the level of this deck for passengers to embark, and if so, how the passengers were to climb over the side into the boats without falling into the water, I don't know. And I doubt if any other passengers knew. As far as I could judge, the sole result of the boat-drill, for the passengers, was to teach them how to put on their life-belts (an absolutely fool-proof operation), and to familiarize them with their boat-stations, which they must have known already, the same being plainly billed in every cabin thus: "*Your boat-station is No. . . . starboard* (or port). In the event of etc., etc., put on your life-belt and stand by at your boat-station. Boat-station numbers are painted on deck. (So they were, very conspicuously.) It is your duty to know the location of your boat-station and life-belt." I hope that no maritime mishap will occur to this splendid ship.

It is remarkable how traditions survive at sea. In this ship meals are announced twice everywhere by bugles—half an hour or an hour before the time, and a few minutes before the time, of serving. The first bugle sounds at 8 a.m. No horn of elfland faintly blowing, but a fearsome, strident, brassy row, which must infallibly waken everyone who is asleep. The bugling, entirely unnecessary, implies that the Command denies the right of any passenger to sleep after 8 a.m.—on a holiday too. Or it implies a complete lack of imagination on the part of the mandarins of the Line.

London.

I had lunch with a young woman aged three. After a perfect dickens of a row she had been sent to bed supperless last night for obstreperous and obstinate naughtiness. This affair with its horrid sequel must have provided her with sensations quite novel in her experience of the world. To-day she exhibited the manners of a Count d'Orsay, the sweetness of a Louise de la Vallière, and the reasonableness of a John Stuart Mill. And yet there are parents who maintain that the best way to teach righteousness to children is by example only, never by correction or deprivation.

London.

A man's dinner in the house of an epicure. The principal dish was cold lamb. At an ordinary 'mixed' dinner no host or hostess would dare to offer cold lamb. But this cold lamb was amazing and amazingly sweet. Its reception amply justified the host's courage. Conversation general. Old friends argued abusively in perfect contentment and security. Consumption of alcohol very moderate indeed. We broke up early. If all dinner-parties were as satisfactory—!

London.

Some people have odd notions of romance. I have. When the dailiness of the domestic atmophere oppresses me in the afternoon, and what I am writing seems flat and getting flatter, I sneak out of the house and go to a Dairy Company's tea-shop in King's Road, and order a pot of China tea and read "Les Nouvelles Littéraires," and neglect "Les Nouvelles Littéraires" in order furtively to watch my fellow-wassailers. A tea-shop is a wonderful place, in a high-degree romantic. The people who come in and go out. The girls together, feminine. The men together, masculine. The men who would no more think of taking their hats off in a tea-shop than in a railway station. The different tones in which orders are given. The different demeanours which the same attendant will adopt towards different customers. The staid supervisors. The slatternly dish-washers peeping forth now and then. The cash-girl eternally in her cage, and in the draught from the ever-opening and ever-shutting door. One talks of the romantic quality of Paris cafés. They are prose compared to the free-verse of a London tea-shop.

London.

Nursemaids. The mother of an infant related to me the following. She had a tremendous scene with the nurse. The occasion, but not the cause, of the scene was the fact that the nurse had kept the nursery too cold, and had then completely undressed the invaluable infant before having made the preparations for its bath. So that the child was kept for at least five minutes naked in a chilly room. Vituperation flew from one side and insolence from the other at a great rate. (The mother laughed as she told me the story in detail.) The mother then began a sentence: "I do hope, nurse, you won't——" "Oh, you needn't fear!" the nurse quickly interrupted her, "*I shan't take my revenge on the baby when I've got him alone.*" Dramatic revelation of a mentality! The mother was only going to say: "I do hope you won't let this occur again before you leave."

Two days earlier than the above another friend who knew Sweden and had many friends there told me that baby-nursing is in the curriculum of *all* Swedish schools, so that every girl, aristocrat or plebeian, acquires experience of all branches of baby-minding. The swells practise on the babies of the peasants, and carry them out for walks in the street for promenades.

Moscow, August 28th.

I went to the opera, and saw an Act of a most tedious ballet, alleged to be modern, but in which I could perceive nothing but a futile spirit of reaction. My new novel, which I have been on the edge of commencing for a year past—without having commenced it, was worrying me into a fever of apprehension. My whole future seemed, and seems, to depend on the quality of that novel. I had the idea for it years ago. I saw the thing vague, but magnificent, tremendous, the greatest novel that ever was or could be written by anybody. Then I lost it. I mean I lost the creative mood for it, and couldn't regain the mood. Then, later, I began to see the thing afresh. I pieced two plots together and made one. I saw the chief characters, and the chief incidents, and the climax. The trouble was, I still could not regain the creative mood for that particular book. I saw, but didn't feel. Everything was there except the breath of emotional life. The spectacle of the ballet extinguished in my inefficient noddle the last glimmer of hope.

The next item on the programme was an Act of Rimsky-Korsakoff's opera, " Sadko ". I'd never seen " Sadko ", and I doubt if I had met anyone who had seen it. Anyhow, I knew Rimsky was not a really first-rate composer, only an agreeable melodist and a terrific swell at orchestration. I expected little from " Sadko ". But I had a surprise. The music was so close to being first rate that I was unable to tell the difference, and the performance was marvellously fine. Something of the old autocracy had survived into the Soviet autocracy. Before the Act was half over, my novel was coming back to me in quite the grand manner. I could listen to the opera and think about the novel simultaneously. I felt the creative mood permeating and enveloping me. Hope returned, higher than ever. At the end I applauded with the enthusiasm of a youth. " Sadko " was my salvation. I said to myself joyously: " Well, I haven't come to Moscow for nothing."

And as soon as I was outside the theatre I knew I had got hold of the affair, because everything that caught my attention related itself to the novel, gave me fresh notions for the novel. Which is a sure sign to an artist that God's in his heaven. I have often been through a similar experience, but perhaps never quite so dramatically and suddenly. A novel in process of creation has to be lifted up.

It may have to be lifted up again and again. The large mood for it has to be recaptured again and again. To work this miracle there is nothing as efficacious as the sight or hearing of a great work of art—any art. Many times have I gone into the National Gallery, or to a fine concert, not primarily to see pictures or to hear music, but to recover the right mood. An artist engaged on a work ought never to read or see or hear second-class stuff. If he does, he realizes the resemblances between his work and the second class; and is discouraged. Whereas if he sticks to first-class stuff, he realizes the resemblances between his work and it, and is enheartened thereby. I've felt this a thousand times, and said it a hundred times, and perhaps I've written it ten times. But it can't be repeated too much.

Danzig, August.

The broad avenue between this city and one of its pleasure-suburbs has two roadways separated by a continuous lawn of well-tended green grass. I was startled to see a tram running over this lawn. Then only did I notice that there was a double set of tramlines between the two roadways, and that grass had been grown between and on either side of the rails. This device made all the difference to the appearance of the avenue. The desert is arid, but less arid than a double set of gravelled tramlines, and not more dusty. Some one had thought of the device. He possessed imagination.

And the city council, his employers, had the wit to consent to bear the slight expense of his imagination. Rare.

Oslo.

Only when you see the provincialism of Oslo do you appreciate the wonderfulness of Ibsen. It was as manager of the theatre at Bergen that Ibsen learnt for himself more about the stage than any other dramatist in nineteenth-century Europe. Now Oslo is a capital; it has three times the population of Bergen; it is much nearer the cosmopolitanism of Sweden. If Oslo is provincial—and you may even call it parochial—to-day, what must Bergen have been like in those early years when Ibsen, directing its theatre, formed his ideas and planned his schemes for the rejuvenation of the drama? Whence came the inspiration which enabled him to make all the plays of the continent seem petty, parochial, and ingenuous in comparison with his own? This is a mystery which cannot be explained, and certainly Ibsen himself could not explain it. I doubt whether any creative artist ever can satisfactorily explain his causation.

I remember the time when Clement Scott in the *Daily Telegraph* used to attack Ibsen violently, to shoot him to pieces with epithets. (Ibsen never seemed to notice that he had been shot to pieces by the most influential dramatic critic in the biggest city in the world.) The mildest of the charges brought by the angered Scott against Ibsen was that he was parochial. In those days, thirty or forty years ago, I was indignantly anti-Scott. Ibsen parochial! The notion was grotesque. But to-day I do have a glimmering of what Scott was driving at. In a way Ibsen *is* parochial. (So was Aeschylus, and so was Thucydides.) It was like Ibsen's immense cheek to assume that the élite of Europe could be interested in the back-chat and the municipal and connubial goings-on in a twopenny town of a sort that nobody had ever heard of. Ibsen's assumption nevertheless proved to be correct.

Yes. Ibsen was parochial, even in his finest plays of contemporary life; but he lifted parochialism to the mundane and the universal. Read or see Ibsen's social dramas without prejudice, and the still small voice within you will say: "But I know that town and its inhabitants. I have lived in it, and among them. I *am* living in it." Fundamentally, we are all living in Bergen. Ibsen knew that Bergen was everywhere. He was a genius.

Not that the word "genius" explains anything! It soothes. If a not-too-sick man sends for a doctor, and the doctor diagnoses his malady, the patient is soothed, because he thinks he has learnt something. He has learnt only a name. Similarly with all of us when, puzzled past bearing by works of a creative artist, we call the fellow a genius! We soothe ourselves with a name. But we are no forwarder.

One symptom of the provincialism of Oslo is the way in which people talk to one another in the street. Watch a chauffeur ask a policeman for directions about a route. Watch the policeman's reply. The chauffeur and the policeman are alike both brusque and chatty. Their tones are clipped, their syllables dropped towards the ground, negligently, as though the speaker didn't care a farthing whether anyone picked them up or not. (You observe the same thing in shops and cafés.) The conversation has apparently none of the juiciness of the metropolitan, and yet every now and then you perceive the transient gleam of a drop of true human juiciness. The people have a northern instinct against admitting that a cocoanut is not the same inside as out.

It is said in England that all Norwegians can speak English. Nearly all the Norwegians whom you are likely to meet do talk English. But for the most part it is Norwegian English, clipped and then curiously shaped into sentences. Often you are not quite sure what the amiable Norwegian means. Still more often, the Norwegian is not quite sure what you mean. I heard a man say to the ticket-seller at a museum: "Can you tell me how long this museum has been open?" "Yes," she answered. "Till three o'clock in the afternoon." Withal, Norwegian English, if elementary, is exceedingly useful, and the travelling English ought to thank heaven for it, and to inquire at what school in England, if any, a boy can learn to speak any foreign tongue as well as a Norwegian speaks English.

London.

Eight years ago I bought a portrait of a woman by Modigliani—certainly one of the greatest painters of this century—for £50. So that when I received an invitation to a private view of Modiglianis in a West End Gallery, I accepted it at once. There were no £50 items in this show. I halted before the picture which pleased me most, and asked the price of it. The manager replied: "A Paris dealer offered me £6,000, but I refused it." This news delighted me. He did not say what his own price actually was, and I did not inquire further. I never paid more than £100 for a picture in my life, and I never will. At this private view a professional photographer came up to me and asked permission to photograph me in the act of gazing at the masterpieces of Modigliani! I forbade. He then went up to a well-known collecting peer, and made the same request, and was again rebuffed.

London.

I dined with a small group of economists, men of science, ethnologists, etc. Most of them bore illustrious names. Their various learning, and the quickness and freedom of their brains, intimidated me. So I assumed the safe rôle of listener. Towards the end, one of the most illustrious names, aged about 66, said to me: "Do you dance?"

I said: "Yes." He said: "Well, let's go to the —— and have a whiskey, and then I'll take you to a dance-club. We are certain to find partners." I said that I never drank whiskey, and that I must go to bed early, as I had a full programme of work to-morrow. He gazed at me, and his vivacious eyes said: "Slave!"

London.

A journalist related to a financier in my presence how his stockbroker had advised him to sell shares in an American utility company at 18*s*. and how, very soon after he had sold them, the shares rose to 24*s*. 6*d*. The journalist desired to know how this expensive kind of advice could be explained. The financier replied: "The explanation is quite simple. You merely have to remember that every stockbroker—even the best stockbroker—always wants his customers to do two things—to buy shares and to sell shares, and it doesn't matter much to the stockbroker which of the two things the customer does." The financier added that the majority of Stock Exchange transactions were for only about £300 worth of stuff.

London.

Talking familiarly at lunch with a London police-magistrate, I asked him: "Do you really like your work?" He said: "I love it. I'm never bored by it. I shall be at it even this afternoon, much to the annoyance of counsel, who would prefer to be watching the race for the Schneider Trophy. It was always my ambition to be a police-magistrate, and I'm very happy in the job."

"But," I said, "isn't it a bit monotonous? Same sort of cases recurring again and again, and so on."

He said: "You might think so. But the fact is, no two cases are alike, and no two cases have to be treated in the same way. There's no monotony in a police-court,—at least there isn't for me. A police-court is all human nature, and I haven't yet found that human nature is monotonous. Now yesterday I had a case. Two old gentlemen, aged sixty-five and seventy, quarrelling about a waistcoat valued by A at ten shillings, and handed over by A to B for safe keeping. Both these gentlemen resided in a Rowton lodging-house, and earned a precarious living by addressing envelopes. A demanded the return of the waistcoat and B wouldn't part. B made excuse after excuse, until A came into court with his grievance, demanding the waistcoat, or in the alternative ten shillings cash. I discovered that B has been wearing the waistcoat —no doubt he had none of his own—and that it was no longer absolutely clean. I decided that the waistcoat was worth five shillings, not ten, and I ordered that the waistcoat should be handed over within a week, or five shillings paid. A then said to me: 'I hoped your worship would make a specific order for payment in cash, because, as the waistcoat has been in wear by Mr. B since April, it would not be satisfactory to me to have a *restitutio in integrum.*' "

A tragedy hidden somewhere behind those last three words!

About the Schneider Trophy race. I looked at half a dozen morning papers to discover the time of the start. In vain. Every particle of information was given except just that particle. I mentioned this strange fact to the magistrate, who at once said: 'As the race is to be broadcast, you can deduce the time from the Wireless programme.' Ingenious! I did. Then I was conscious of a sudden desire to be down Portsmouth way. Not because I wanted to see the race, nor to hear the roar of it. But because the papers had been making such a tremendous fuss about it, and because a million of my fellow-creatures would be gathered together to watch it. The herd-instinct again! Perhaps no one is entirely free of it.

London, September.

I went by invitation to the "world-première" of an English-written and English-directed talking film, in which Gloria Swanson was the star. The film was apparently made in America. My opinion of Gloria Swanson's gifts as an actress in silent films is very high indeed. I was bidden for nine o'clock, and at nine o'clock I arrived.

The street in front of the theatre was crowded with sightseers, some of whom were perched on the tops of lorries used as grandstands. A broad path across the pavement was kept clear by the united efforts of policemen and theatre officials. As I passed between the stalwarts I was the subject of loud remarks from the populace. The big theatre was crowded, except in the best seats round about me, which had been reserved for guests whose names have a publicity value. Many of these empty seats were never occupied during the evening. A silent film was already in progress, and it continued in progress for an hour or so. What qualities it had to recommend itself to my attention I failed to see. However, it did at length finish. Then a gentleman came in front of the curtain and said, *inter alia* : " Miss Gloria Swanson is in the audience and if you will kindly remain in your seats for one minute after the conclusion of the new film, you will see her." At these words there was a great noise from the audience—a curious kind of clapping not intended to signify approval. The talking film began. The noise increased. So much so that the film, though it could be seen, could not be heard at all. The film-operator and the audience were equally obstinate for a minute or two. The audience won. Gloria Swanson, who was seated a few rows behind me, stood up in the gangway and bowed. Useless! Half the audience could not see her. The audience grew still more restive. The noise was resentful and imperious. It seemed to say : " She belongs to us. She is ours by right. Show her."

She left the circle, and was presently seen walking

up the central aisle of the floor, well escorted. Then she came before the curtain, obviously in a highly nervous condition, and made a little speech, which was almost inaudible. As soon as she had retired, at least two-thirds of the huge audience on the floor stood up and hurried from the theatre. They had come to see, not the film, but Gloria Swanson. Having seen her, they departed. Surely rather odd.

The film started again, to many hundreds of empty seats. I could discover no originality whatever in the film, and no merit except the striking merit of Gloria Swanson's performance. The story somewhat resembled that of "East Lynne"; but it was not as good as "East Lynne". Crude, tawdry, grossly sentimental, encumbered with stretches of acutely tedious and undramatic dialogue, and rendered ugly by the continuous falsification of the sound of the human voice which mars all talking films, it crawled along from foreseen crisis to foreseen crisis in the most exasperating manner. Its attempts to be noble were merely distressing.

But Gloria Swanson was magnificent in it. She proved that a great star of the silent can be equally great as a star of the talking. She used extreme technical skill, and displayed throughout both real power and real distinction. She even sang. The songs were her one mistake. The film did not demand song, and her singing was amateurish At the close she appeared once more before the curtain and made another little inaudible speech.

I left the theatre saddened by this spectacle of the waste of a first-rate artist. The space across the pavement was still being kept by policemen and commissionaires. The crowd was larger than before, but order was being maintained. Then suddenly order vanished. The two lines of stalwarts were smashed in an instant, and I was being tossed to and fro in a mass of hysterical women. Gloria Swanson had appeared in the entrance-hall. She fled back. I gave a stalwart one shilling to act as a spear-head for my party through the wild surge. He was not overpaid. In ten seconds we had reached safety. Cries! Shouts! Shrieks! Clapping! Order was restored and Gloria Swanson slipped into the film-star's immense and luxurious automobile which was waiting for her. What an evening! What a light thrown on the mentality of the film-fan! I restrained my sympathy for Gloria Swanson. She is a queen-empress. She does what she chooses. She is a woman of experience, and she must have known what she was in for.

London, September.

I think I must have been out of England when a suggestion was seriously put forward for making it illegal to build automobiles capable of travelling at a speed greater than twenty miles an hour. Anyhow, I only heard of this brilliant scheme the day before yesterday. The author of it was a man of sixty, celebrated himself, and bearing a name very celebrated in the political history of England. It really astonished me. Yet I thought I had long got beyond the possibility of being astonished by any vagary of a politician.

Here was a man accustomed to political life, once a Minister of the Crown, a respected figure in Parliament, a man surely not unacquainted with the nature and force of public opinion, giving away the true quality of his brain by a suggestion of the most enormous fatuity, and not even suspecting its fatuity. Perhaps it was a trifle that his project, if realized, would have the result of increasing the number of motor accidents, which are avoided as often by a burst of speed as by an application of brakes. Perhaps it was a trifle that his project would kill completely the export of British motor-cars.

It was not, however, a trifle that this man, whose moral and intellectual standards have been admired even by his opponents nearly as much as by his friends, should seek to enforce a speed-limit not by moral appeal but by mechanics. Nor was it a trifle that he should prove himself to be pathetically ignorant of the British character. Of all civilized peoples the British are the most law-abiding—if the law shows any glimmering of reasonableness. On the other hand, if the law loses all reasonableness, the British people *en masse* will defy it and no efforts of policing and punishment will persuade them to obey it. The twenty-mile speed-limit has long since become absurd.

Not a single motorist on the road conforms to it. It is the object of universal derision. It is a farce. And yet this reputable man, all other means of enforcement having failed, would enforce it by automatic contrivance! Does he imagine that he is living in the twelfth century instead of the twentieth? Since when have British citizens deserved to be treated like children? Has any government tried to treat us like children without incurring either a Falstaffian laughter or a catastrophe? True, the project was utterly silly, and as a project was without any practical interest. What was important about it was its revelation of a mentality, unimaginative, ignorant of fundamentals, bereft of common sense, and disastrously wrong-headed. The mentality of a politician who in the past has exercised influence in the legislative organism, and will probably exercise it again! And be sure that the politician in question is not the only one of this calibre. There are others, possibly many others. It was this aspect of the egregious episode that frightened me.

London.

I was finishing up a great evening-waistcoat enterprise at the tailor's to-day. He is far more of a hosier than a tailor. He told me that he employed a man who did nothing year in year out but cut and superintend the making of white waistcoats. Some people might regard this as rather a narrow career for an immortal soul. But it is infinitely more complex, subtle, and broad than the career of a man who sprays paint on to the bonnet of a popular motor-car and is expected to cover, and does cover, three hundred bonnets a day.

London, September 25th.

To-day I began a long novel. At 3.30 p.m. The hour has an interest—but only for me. I have not written a long novel for years. As a man with a secret tendency towards idleness I prefer to write a short novel. It is easier. Not easier to *do*, but less of a strain on the creative faculty. What generally spoils long novels is the untimely supervening creative fatigue. This happened to "The Heart of Midlothian". It is a calamity which the author has very little power to prevent. Heaven in its wisdom decided to give you a certain amount of strength. You cannot increase it. Towards the end of a race, if you are tired you are tired, and there you are! Nobody can pour a quart out of a pint pot.

Lots of good novels fall away a few score pages before the end. Readers say: "The author scamped this last bit." He usually didn't. He was at the end of his creative strength. He may have had, and probably did still have, reserves of invention, ingenuity, perseverance and conscientiousness; but his *power* was exhausted. He had been guilty of only one artistic sin, the sin of miscalculating his creative strength. No work on earth is more trying than creative writing. As you write the first words you are self-conscious. When you finish the first page of manuscript you think: "I have so-many hundred more pages to write. Every one of them has to be written, and every one of them must be good. Every one of them must be the best. No letting down."

I reckon that this novel will fill 900 pages of manuscript. How do I reckon? I don't reckon. I just know. Experience has taught me pre-knowledge. When I began "The Old Wives' Tale", I announced to the domestic hearth: "This novel will be 200,000 words long, divided into four equal parts." Well, it was. The new novel will be 150,000 words long, and probably not divided into parts. I think I have now grown out of dividing novels into parts. To-day such a division strikes me as being a bit pompous. I know the main plot, but by no means all the incidents thereof, though I have a few titbits of episodes which I shall not omit. The episode of the gloves, for instance, which I found in and appropriated from the *Journal* of the brothers de Goucourt. I know the three chief characters, but by no means all the ins-and-outs of them. They won't alter—I would never allow any character to get the whip-hand of me—but I shall fill them out. I know the 'feel' of the novel. That won't alter, either. And I have the whole of the material for the novel; and it is indexed, in a notebook. I would sooner lose fifty pages of the manuscript than that note-book. If I did lose it, I think I should be capable of abandoning the novel for ever. And yet I leave the notebook lying about.

I have been fighting for years against the instinct to write this particular novel. About thirty years ago I was taken to the Savoy Hotel for tea, came out, went home, and wrote " The Grand Babylon Hotel " in three weeks of evening-work. " The Grand Babylon Hotel " was a mere lark. The big hotel de luxe is a very serious organization; it is in my opinion a unique subject for a serious novel; it is stuffed with human nature of extremely various kinds. The subject is characteristic of the age; it is as modern as the morning's milk; it is tremendous, and worthy of tremendous handling. I dare say it's beyond me. But nobody else has caught hold of it, and if I am not audacious I'm nothing. To-day I wrote three pages. 897 left to do! The thought is terrifying. Any serious novelist will agree with me as to the terrifyingness.

And when I have finished it and corrected the manuscript and corrected the typescript and corrected the slip-proofs and corrected the page-proofs, and it is published, half the assessors and appraisers in Britain and America will say: " Why doesn't he give us another ' Old Wives' Tale '? " I have written between seventy and eighty books. But also I have only written four: " The Old Wives' Tale ", " The Card ", " Clayhanger " and " Riceyman Steps ". All the others are made a reproach to me because they are neither " The Old Wives' Tale ", nor " The Card ", nor " Clayhanger", nor "Riceyman Steps". And "Riceyman Steps" would have been made a reproach too, if the servant Elsie had not happened to be a very 'sympathetic' character. Elsie saved "Riceyman Steps" from being called sordid and morbid and all sorts of bad adjectives. As if the 'niceness' of a character had anything to do with the quality of the novel in which it appears! But authors are never satisfied.

London.

A lawyer friend of mine, back from a visit to New York, told me that he had recently been one of a crowd of 47,000 at a prize-fight in Madison Square Garden, where there was ordained an interval for prayer for the repose of the soul of Tex Rickard, the prize-fight organizer, whom everybody present knew to be a great and violent sinner. Jack Dempsey, and a professional toast-master at a terrific salary, stood alone in the ring under a blinding glare of spot-lights, while the whole vast auditorium was darkened. Everybody had to stand with bare *and bowed* head. The professional toast-master prayed. A silence. Then the fighting proceeded.

London.

I lunched near Park Lane, and stopped to talk with an acquaintance outside the dining-room door. We entered an enormous apartment, apparently empty. But after a while I descried in the middle distance a physical object. It was the dining-table with twelve full-sized human beings sitting around it.

London.

To-day I was told that John McCormack the tenor made nearly two million dollars a year by his voice. I said: "He couldn't. If he gave a concert every weekday all through the year and made £1,000 net out of every concert (which would be impossible) even then he wouldn't make nearly two million dollars in a year." The reply was: "You are forgetting his gramophone records. He makes a million and a half dollars a year out of them." I retired leg-before, hurt.

London.

An author of an older generation than mine talked enviously of the royalties of novelists—popular novelists. He was talking 'at' the novelists at the table, including me. He said that his best book had almost no annual sale—a few hundred copies. It was published about thirty years ago. His most popular book was his worst, and last year it had a sale of "only 7,800 copies". "How old is it?" we asked. "Oh!" said he, "I couldn't tell you exactly. It must have been first published twenty-nine or thirty years ago. Perhaps in 1900." These books are text-books.

The novelists present merely smiled. Not ours to give the show away. We might have informed him that the number of modern novelists whose novels reach an annual sale of 7,800 copies after being extant for thirty years is as near zero as makes no matter. We might have informed him that the sale of the ordinary fairly successful novel comes to an end within six months of publication, if not sooner; though of course a small percentage of novels do achieve the cheap-edition stage—a stage, however, which brings but relatively trifling sums to the author.

Cheap editions (3*s.* 6*d.* or 2*s.* 6*d.*) of novels very rarely reach large figures. I doubt whether any novel of mine has had as large a sale at 3*s.* 6*d.* as in its original form, which fact, recurring, always inclines me to doubt the confident assertion of persons with limited incomes that they don't buy novels because modern novels are too dear. Also, it is to be noted that, to my knowledge, two attempts have been made to sell new novels exactly similar in appearance to new novels at three half-crowns, for one half-crown. Both attempts completely failed to justify themselves. Booksellers' shops were not invaded by cohorts of the half-crown public. Indeed the sales were deplorably unsatisfactory. Publishers lost money. So did authors. Insufficient advertising may have had something to do with the disaster. But books at half-a-crown will not 'stand' much advertising. The most popular of all my seventy-four or -five books, published some twenty years ago, has an annual sale of about 3,000 copies, with which I am well content. If it had an annual sale of 7,800 copies, I should be rather more than content. I should be quite puffed up. It is not a novel. It might not improperly be called a text-book, like the book of my senior friend. Its title is "How to live on twenty-four hours a day". I wrote it in a week or two. It appeared serially in a daily paper. And I was strongly advised by an expert not to republish it in book form. I flouted his wisdom.

Yesterday I learned that the writing of text-books for pedagogic institutions is not always remunerative. The working chief of a large business enterprise related to me, in detail, how after a shortish scholastic career he had been engaged by a publisher as general 'editor' of text-books at a salary of £275 per annum. (This was before the war.) 'Editing' the text-books proved to be writing the text-books, in addition to devising all business arrangements. During five years my friend actually wrote entire text-books on all manner of subjects at the rate of one a month—sixty in all. Some of these works still find a more or less regular market. He never got a penny of royalty. He never got any increase of salary. Under a clause in the contract the publishers held all the copyrights. Some publishers are cleverer than others. This particular publisher merited a Nobel Prize for sustained, rock-like cleverness.

London.

At the Gate Theatre I saw the German play, "The Outskirts" (called in the German "Periphery"). At a later date I saw another German play, "Hoppla", at the same theatre. Extraordinary and incomprehensible, the prestige of German plays in advanced London! The Gate Theatre is an enterprise with a genuine artistic aim. It rarely if ever produces a new English play. Its productions are nearly always imported from foreign lands. Both these German plays are ill-constructed or not constructed at all, sentimental, long-winded, full of impossible dialogue, and in a general way feeble and messy. They are not original, but they have an air of originality, because they are episodic, cinematographic, fragmentary, and occupied with low or vicious life. (The same is to be said of the French play "Maya", a series of scenes supposed to depict the existence of prostitutes in Marseilles, which filled a Paris theatre for many months and thereby got a spurious reputation at the Gate Theatre in London.) I saw "Hoppla" in Berlin once, and in the middle of the performance unobtrusively slipped away from the theatre.

I saw a rehearsal of "Periphery" at the Josephstadt Theatre in Vienna, on the invitation of Professor Max Reinhardt, who was producing it there. I believe that this was the first production. The affair had its interest. When I went into the theatre I was received by a member of the staff, who closed even the outer doors after me with the most particular precautions against any noise; for the Herr Professor was conducting a rehearsal! An inner door was closed with similar precautions. I tiptoed into a seat. The auditorium was in darkness, save for one electric light which burned on a temporary desk set in the middle of the stalls. Also on the desk was a telephone. The Professor sat at the desk with an assistant on either side of him. He never spoke, never moved. The telephone was not used. Scene after scene was played without any interruption or criticism from anybody. Then the curtain fell for an interval, and the Professor left his desk and came to talk to me. But he did not talk theatre.

And I have been present at rehearsals of " The Miracle " at Salzburg directed by Professor Reinhardt; and here too he sat still and watched in silence—broken perhaps by one or two brief whispers in the ear of an assistant. I don't know how he gets the effects which beyond question he does get. He certainly does not get them by talking. The contrast between his methods and the methods of restless English producers, who rarely leave the actors alone and who walk about like admirals on the quarter-deck in a naval battle, is very marked.

London, October.

I was noticing the faces and the demeanour of the middle-aged or ageing women-shoppers in Sloane Street. It was painful to observe how few of them can use paint and powder with discretion and effectiveness. Some of the lips were dreadful sights. Then the haughty, hard, harsh expression on many of the faces! The sort of expression that says savagely, during strikes: "Shoot them down!" or "They ought to be *made* to work!" or "Unions ought to be made illegal." A large percentage of these ladies must surely be rather unpleasant to live with. You seldom see such expressions as theirs on the faces of men. I suppose this is because men go about more, and arrive at a notion of the real facts of existence.

As regards politics and industry, a lot of these women are living in the 1880's. They are probably one of the origins of the servant problem. They don't realize that they have nothing whatever to be haughty or self-complacent about. Most of them have done nothing for society at large, and little for the other immortal souls who menially serve them: though of course all or nearly all of them have patronized the poor. The truth is that Sloane Street during the shopping hours is not an entirely agreeable spectacle. I much prefer Oxford Street, where the social salt of the earth do not occupy the pavements. I prefer even Bond Street, which is more international than Sloane Street and less conscious of a sublime superiority. But for the man with eyes to see—naturally I mean myself—the most interesting shopping street is the despised Strand. Nevertheless, the Strand is losing its most endearing and picturesque characteristics. It will soon be widened from end to end. What is worse, it will soon consist solely of 'edifices', and the last of the high, narrow shop-houses will have vanished. And what is even worse, it will soon be tidy. Withal there are but few frills on the Strand—yet.

London.
I read the last 200 pages of Gaboriau's "L'Affaire Lerouge" this afternoon. I must have read it first as a boy, in English. My previous last perusal of the original French was in 1916. Now I have finished it again. I read it in order to compare some new detective-novels with an admitted masterpiece in that kind. There never was such a rage for detective-novels as to-day. Our chief truly literary monthly, written mainly by highbrows for highbrows, gives many pages each month to the appraisal of the latest detective-fiction. "L'Affaire Lerouge" is out of sight better than the British favourites of the hour. I should like to think otherwise, but I can't. It is better because it has a firmer grasp of plot and much more creative imagination.

The first 200 pages of "L'Affaire Lerouge"—the opening, the discovery of the crime, and the preliminary detective-work—are simply brilliant. No flaw in them. But the novel as a whole is by no means perfect. Gaboriau even in his best novels has two faults. His use of coincidence amounts to impudence. There are two terrific coincidences in "L'Affaire Lerouge". One is that Tabaret the amateur detective happens to live in the same house as a group of people intimately and vitally connected with the crime! The other coincidence is that Daburon, the examining magistrate, happens also to be connected with the above-mentioned small group of people.

These enormities would ruin any fairly good detective-novel. But Gaboriau, by his immense force of creative imagination, carries them triumphantly off. He compels you to say to yourself: "Well, it was extremely odd that things fell out so, and extremely convenient for the author, but they *did* fall out so. And that's all there is to it."

London, October 21st.

March of Time. I was bidden to the opening of the new Savoy Theatre—" The Gondoliers ". But I was bidden also to a dinner given by Frederick Lonsdale to some forty friends at a club, before the first performance of his new comedy, " Canaries Sometimes Sing ". When the old Savoy Theatre was new, Frederick Lonsdale was less than a year old. He established himself as a writer of the 'books' of musical comedies—tremendously successful and remunerative. Then he turned to drawing-room comedies without music, and became still more successful and still more handsomely remunerated. I remember my first experience of these non-musical comedies, " The Fake ". It contained one scene, in a hotel, between a man and a chamber-maid, which was masterly in its handling of dialogue, and which gave a glimpse of what Lonsdale was soon to be. I remember thinking with a thrill: " Here is a man who can write dialogue which is both lifelike and humorous." In due course he was producing whole comedies as good as that single scene in " The Fake ".

Now his position is so assured that without too much nervousness he can preside at a large dinner one hour before the curtain rises on the opening performance of one of his own comedies. Audacious proceeding! But, anyhow, less audacious than issuing invitations to a supper *after* a first performance. I have assisted at two after-performance suppers given by dramatists on first-nights. In both cases the play had been a failure. Imagine the atmosphere of the party meant to be festive, jolly! Imagine the lies, white and black, that the author had to listen to!

I have had many first-nights, perhaps nearly as many as Lonsdale; but never have I arranged a party for them, either before or after the performance. Let a dramatist be as aloof, indifferent, successful, godlike as you choose, the strain upon him of a first-night is acute. Experience has taught me to lie low on a first-night of my own. I have attended some of them, and always regretted it. My practice now is to go to a music-hall, or to another play, or to the opera. On the first-night of " Sacred and Profane Love ", I witnessed Verdi's " Falstaff " at Covent Garden. Then I strolled along down to the scene of " Sacred and Profane Love ". I was in no hurry. I could wait. I was determined not to arrive at the theatre until the play was over. But I miscalculated. The performance was late, and the curtain had only just fallen on the last Act when I innocently penetrated into the wings. Noises of extreme enthusiasm! Basil Dean demanded that I should 'take the call'. I refused. I said that nothing would induce me to take the call. The curtain rose and fell, rose and fell, rose and fell. And then Basil Dean, as furious as myself, but stronger, dragged me on to the stage. "Sacred and Profane Love" failed in London. Fortunately for me it was a considerable commercial success in the United States. But then I had not put the evil eye on it by appearing at the first performance in America.

From Lonsdale's party I proceeded to the new Savoy, the most modern theatre in London, with decorations that will certainly set a fashion. All London was of course at Lonsdale's first-night, but all London was equally at the opening of the new Savoy. The fact is that all London could easily fill half a dozen theatres. Nearly fifty years since the old Savoy Theatre was opened! And the old Savoy was once the most modern theatre in London, so modern that many judicious people feared that it might be too modern. It was the first theatre to be entirely lighted by electricity. (But the cautious D'Oyly Carte had a gas-installation in reserve, for he also felt qualms.) It was the first theatre to have queues. It was the first theatre to have programmes in the form in which we now know them. And a few months ago this once-too-modern theatre was ripped to pieces and removed from within its four walls because it was regarded as too seriously behind the times!

Not everybody realizes that the Savoy operas were the first musical comedies. They were, though. Fanciful plot. Love interest. Comedians. Concerted numbers. Chorus girls. And all. The only difference between, say, "The Gondoliers" and any modern musical comedy is that the former is in every way better than the latter. The one non-Savoy musical comedy that comes anywhere near the Savoy pioneers is "The Chocolate Soldier", and for the plot of that Bernard Shaw was responsible! Compare "The Gondoliers" with the most popular musical comedy of the age. "The Merry Widow"—with its fun depending on drunkenness and debauchery. Embedded in the wild fantasy of its plot, "The Gondoliers" has social criticism which is as apt to-day as when it was written. "The Gondoliers" has form, balance and pattern. It is as shapely in its artificiality as a play by Marivaux. It is classic.

But this night was Sullivan's, not Gilbert's. The libretto is good, but the music is much superior to the libretto. The outbreak of the first great chorus in the first Act was positively sensational in its power and beauty. And it was really *sung*. No chorus-singing in any modern musical comedy was ever half as musical, as animated, as fine, as efficient as the chorus-singing at the new Savoy to-night. My two organs of vision and my throat were affected. No doubt one reason for emotion was the presence, sitting side by side in the audience, of two Savoy ex-stars: Jessie Bond and Geraldine Ulmar. March of Time!

London.

One of the most celebrated actresses in the whole world came to dinner. She smoked two cigars. She said: "I do like coming here. I adore cigars, and this is the only house where I can smoke them." I said: "But you could smoke them anywhere." She said: "No. It wouldn't do. People would say I was unsexed." And who could deny this? Not I. It is a curious thing that whereas thirty years ago women had to smoke cigarettes in secret because a public cigarette had the magic effect of unsexing them, to-day they can smoke cigarettes openly and keep their womanhood. And yet cigars still retain the unsexing property! And even to-day, in some places, cigarettes continue not to be morally innocuous. A friend told me that in December last he was staying in a country-house and the hostess said to him: "Of course we don't object to smoking in the garden."

London.

This city is the city of strange encounters. I had two to-day. I was walking in a certain street on my way home to lunch, in perfectly plain, fine, unromantic autumn weather, when a young, well-dressed man overtook me and addressed me by name. I can remember all sorts of useless things, such as the number of the house in which I was born and the date of the coronation of Charlemagne; but I have the disastrous habit, when confronted with a sudden crisis, of entirely forgetting both the names and the faces of people quite well known to me. The young man smiled confidently at me, and I smiled at him.

"Hello!" I said, and, feeling sure that I knew him, I took his arm. Then I said, uncertain: "Who are you?" He said: "Oh, I'm nobody. But I wanted to speak to you. I hope you don't mind." (He did not guess that he was gravely interrupting creative reflections upon my new novel.)

He then informed me of his name, and his age (nineteen), and said further that his father kept a raiment shop close by, and still further that his father, in addition to a retail business for men, owned a manufactory of ladies' underwear. He told me that he had followed my printed advice to keep a Journal and how to live on twenty-four hours a day, and that he had read H. G. Wells's "Short Outline of History" and intended to read H. G. Wells's longer "Outline of History", and that he had written some articles and read a short history of philosophy, and desired to make a thorough study of philosophy, and desired to write but could not decide whether it would be better to start in the lowbrow vein or in the highbrow vein. And would I give him counsel?

He walked a good three-quarters of a mile by my side on chief pavements, and I forbore to tell him that he had snatched my novel out of my mind and cast it into the gutter three-quarters of a mile back. He said he hoped I didn't object to being accosted in the street by a stranger; and, lying like anything, I said that on the contrary I was very interested.

Then he related family differences. His father had taken exception to his habit of giving himself afternoons off from the manufactory in order to study philosophy. He had demanded from his father a room to himself in which to improve his knowledge and to write. His father had refused. He had thereupon left home, determined to earn a living by his pen. A fortnight's experience had convinced him that he could not so soon earn a living by his pen; and he had then returned home. He glossed over the exact details of the meeting with his father. But, anyhow, the return had had a fortunate result for him. His father, doubtless frightened by the revelation of the youth's adventurous disposition, had yielded on the main point and had provided him with an income sufficient to enable him to live away from home in a room by himself and study philosophy and write.

I advised the young man to start his literary career in the lowbrow vein, with articles. When I suggested that, in the matter of articles, the first thing was to make them of the right length for the papers and periodicals aimed at, he said: "Oh yes! I've bought 'The Writers and Artists Year-Book'. That gives all particulars." He did not say so, but I thought that the notion of buying that Year-Book must have come from some didactic article of mine. I then stood still firmly. "Good-bye," I said. "Good-bye. Thanks very much," said he. I venture to call this encounter romantic.

The same evening I took supper at a house which is a meeting-place for all sorts of artists, of both sexes. I met a poet there. He was young; he was modest; he remarked in a somewhat sad tone that I rarely mentioned poetry in my articles on new books. I told him I gave poetry a miss for the good reason that I had no technical knowledge of prosody. (True, you can have a knowledge of prosody without having a feeling for poetry, but you cannot properly assess poetry without knowing a lot about prosody.) This young man appeared to be in a fairly prosperous condition.

He had with him some specimens of his work. I asked to see them. Happily they were printed. So I read them on the spot, and I certainly thought that they had some of the stuff of poetry in them. Each poem was printed on a separate sheet. So far, there was nothing very unusual in the affair. The unusual came when the young poet calmly told me that he went himself from front-door to front-door of houses, selling his poems at *6d.* apiece. "But does anyone actually buy them?"

I asked. "Yes," he answered, "I sell a fair number of them." I always knew that London was full of odd phenomena ; but this was assuredly the oddest thing I ever struck. I singled out one poem as being the best. He neither agreed nor disagreed. He was cautious. He said : "That's the one I sell most of." Which somehow caused me to think that I have been wise not to deal critically with poetry.

London.

Tact. A British film-magnate said to me this afternoon : "I'm using one of your books, 'Lord Raingo', these nights to send me to sleep. You know what I mean. 'Lord Raingo' is my dope just now." He said later : "I'm amusing myself by putting real characters to your characters." I said : "Don't. I implore you not to do it." He said : "Oh ! But that's the only interest of reading the book. That's the amusement. I like Andy Clyth's 'yellow eyes'. I've seen those yellow eyes. And I like his long legs. Oh ! I like the long legs."

London.

A youngish Canadian ex-soldier had become interested in a charming blonde English girl who served in some capacity in a country-house where a friend of mine was staying. So interested that he offered to give her a day's jaunt in London. She accepted. They went. "First-class and everything." Return tickets. In the first-class carriage was a small boy travelling alone. The child cried all the time. The charming blonde took no notice whatever of the child, made no attempt to sympathize with him in any way. The Canadian waited and waited for her to behave to the forlorn child as a kind-hearted woman should. In vain. At Waterloo he said laconically to the charmer: "Here's your return ticket." And walked off and left her. He could not stand a woman like that.

London.

At a banquet. After the toast-master had very noisily rapped for silence, the Chairman rose and began: "My lords and gentlemen——" Invidious phrase! Why should he have been at pains to imply that peers of the realm are not gentlemen?

London.

The feature of social gossip, chitchat, scandal is still growing in the press. I know rather intimately several journalistic gossipers, and I never hide from them my dislike of the feature. (Which dislike does not prevent me from reading the feature when I happen to see it,—there are two persons in all of us, even in the righteous and the self-righteous, one who on principle objects to an evil, and another who in practice often yields to the temptation of the same evil.) I have entreated the gossipers never to refer to me in their string of paragraphs. But they frequently do. When I upbraid them, they reply that they have to live, and that copy is copy and much more important than promises.

This social gossip feature is supposed to make a powerful appeal to the popular public. It may: it probably does. But not always. I recently mentioned in an article the name of a well-known very highbrow literary critic. What is more, I praised him. A domestic servant in a certain house happened to read my article, and she was particularly interested in it because the eminent critic lived next door. Now the morning after my article appeared she was cleaning her front steps when the critic came out of his house. Looking at him through the railing which divided the front-steps of the two houses, she said sharply: "You are an ass."

The psychology producing this remark is really rather difficult. I surmise that the girl must have misunderstood the tenour of my remarks, and that anyhow she was averse to publicity for a man who lived so close and who she thought was not properly entitled to publicity. Still more strange, she proudly told her mistress what she had said to the critic. The mistress, although she was not acquainted with the family of the eminent critic, called on his wife and apologized.

London.

I have always regretted that I did not go to a University. I regret the friends and acquaintances that I should probably have acquired there; and I regret even the knowledge that possibly I might have acquired. But there are advantages in not having been to a University. I discovered one of them to-day at lunch when a publicist of sixty, who had graduated at Balliol, told me that 'in his day' it was compulsory to learn the Thirty Nine Articles *by heart*. (He mentioned incidentally that the Thirty Nine Articles contain some very fine examples of English.) He said: "They [the examiners, I presume] would put you on anywhere in the Thirty Nine Articles, and if you fell down, woe to you!"

Another Oxford man, younger than the first, said that in his day the Journeys of St. Paul were of the highest importance, and that any weakness in the recital of them meant disaster. In the case of one otherwise extremely brilliant man, an imperfect acquaintance with those Pauline travels did in fact bring about disaster. A Roman Catholic lawyer, talking afterwards to the victim, said, instead of condoling with him: "Haven't we been warning you for hundreds of years against the danger of reading the Bible?"

The first of the above-mentioned two Oxford men related to me some of the circumstances which caused the dismissal of a curate in a north-midland parish. The vicar, being in serious need of a holiday, put the curate in charge of the parish and went away. The congregation at the beginning of the vicar's holiday was a steady forty or fifty substantial and respectable persons. When the vicar returned he found his church crammed with six hundred working men, and not a single substantial, respectable parishioner in the place. Soon afterwards the vicar received a letter from a wealthy parishioner, who was a brewer. The brewer had promised to give a new organ to the church. The letter contained details of a sermon preached by the curate on the evils of alcohol and the responsibility of brewers. The writer said: "Either the man who preached that sermon goes, or I withdraw my gift of a new organ." The curate went.

London, November 13th.

On Sunday last I attended the special performance for V.C.s of "Journey's End". The V.C. decoration is worn "before all others", and evidently the prestige of the V.C. stands before all others; indeed I should say that it surpasses all other prestiges of decorations put together. What crowds and what policemen outside the theatre! Of course not such crowds and not so many policemen as would embellish the arrival of a Hollywood film-star at a cinema-theatre. No, no! We must not expect too much from ourselves. But I doubt whether any collection of K.B.E.s, K.G.s, K.P.s, G.C.I.E.s, etc., would draw a quarter of the multitude that the V.C.s drew. And I am quite sure that a collection of O.M.s would not draw five per cent of the V.C. multitude.

Well, there are only two decorations free from some kind of stain of undeservingness. One is the O.M. and the other is the V.C. And I have doubts about even the O.M. The V.C. is the reward of " conspicuous bravery in presence of the enemy ". It is nearly seventy-five years old. And less than ten years have elapsed since somebody in authority had the idea of extending it to women. Nurses had on occasion been conspicuously brave in presence of the enemy for more than sixty years while the greatest of all decorations was denied to their heroism!

In the theatre V.C.s were all over the place—hundreds of them. Most of them wore three to six medals; but a few had only the crimson ribbon and the bronze Maltese Cross of the V.C. Without the ribbon and the cross nobody could possibly have picked them out as heroes, though they were in the main a hefty and rather challenging lot. And they were unconventional, careless in demeanour, self-unconscious, and the very reverse of dandies. The large majority seemed to be young and youngish men of strictly limited incomes. We give our V.C. heroes £10 per annum, and if they conspicuously risk their lives a second time, we give them an extra £5 for every new peril. In special circumstances we even increase their allowance to the grandiose amount of £50 per annum. There is nothing like reckless generosity. It proves a bounteous soul.

The younger men excited admiring respect in

me. But the old men, the ageing men, the portly men who had in the worldly sense 'succeeded', all the mature men with their dignity of presence,—these excited in me more than admiring respect. They excited in me the tender feeling which pathos excites. They had once been young and adventurous and audacious. They had done marvels of audacity; but all that happened a long time ago, and now they were sobered and a bit prim, and conscious of experience, and deliberate in gesture. I say it was touching. A strange night, impressive, disturbing to one's sense of relative values! Why do we put physical courage before all other qualities? I don't know. But we do. And since we do, there must be some reason for the preference deep in the primal instincts of human nature. I could argue that moral courage is rarer, and may be more intense, and may certainly be richer in good results for society, than physical courage. But if a decoration was instituted to bestow upon the doers of conspicuous acts of moral courage in the presence of rooted social prejudices, should we seriously think twice about it? We should not. Ten to one we should laugh at it.

London.

A day or two ago, talk with a magistrate and two barristers about a police-court case. A girl had married, and her husband had very quickly deserted her. She took to drink. Then she took to the streets. Then she quarrelled with another man, whom two of her friends amiably held down on the pavement while she kicked his face in. This incident brought her into the police-court. The magistrate, to use his own phrase, regarded her as "more sinned against than sinning". (But about that I should have liked to know—what the magistrate did not know—why the husband deserted her!) And also she was so young—twenty-two. The magistrate gave her the choice between prison and a Home. She chose the Home. She entered the Home, but the next morning at an early hour she walked out of it. Less than twenty-four hours of the Home had sufficed for her free spirit. The police were now after her again. In due course she would be before the magistrate again for a sentence.

The question was: What ought the sentence to be? Three months? Six months? 'Modified Borstal'? No one present could make a useful suggestion. The girl appeared to be doomed.

The conversation shifted to criminal trials. Apropos of the recent decision that in criminal trials the Crown counsel should in future be deprived of his traditional right of the last word to the jury, on the ground that the last word gave him an unfair advantage with the jury, both the barristers were strongly of opinion that Crown counsel should lose also another right—the right to open the case with a formal presentation of the facts of the alleged crime as the Crown saw them and wished the jury to see them. The barristers held that the opening speech for the prosecution was at least as unfairly advantageous to the Crown as the final speech could be. It coloured the minds of the jury from the start, impregnating them with the idea that the accused was guilty, and no amount of rebutting evidence could entirely do away with the effect of the opening speech. The barristers, and the magistrate also, were agreed that a criminal trial ought to begin, without any forensic and tendencious preliminary, with the evidence for the prosecution. They said that such was the method in Scottish criminal courts.

It became clear from the conversation that English legal procedure puts a very heavy handicap on any prisoner, and that a prisoner without professional help has almost no chance of being acquitted. Lastly, none of the three lawyers was convinced that the procedure of British criminal justice was in practice any fairer to the accused than the French procedure which so often rouses the indignation of Englishmen. And they would not deny that it was even less fair!

London.

I had a long talk yesterday afternoon with the mother of a girl graduate at Cambridge and of a boy high up in a public school. In such conversations one does get at a few facts. Both girl and boy have somewhat unusual intelligence, but they are young people with the usual instincts of youth, in no way abnormal, precious or priggish. The girl is interested in political work. At the age of nineteen she could take an active part in an election and even make a speech. On the eve of an election in which she was helping, in the midst of all the turmoil of the last day of canvassing and speechifying, she came running up to her mother in the committee-rooms of the candidate, and drew her out into the dark street (because there was no other privacy), and in a voice of feverish excitement—apprehension mingled with pride—said to her: "Oh, mummy, what *am* I to do? So-and-so has just proposed to me." So-and-so was a co-worker. Which shows what lovely and astonishing things can happen in the dustiness of an election! So-and-so, by the way, was ultimately answered in the negative. Sad, perhaps, but anyhow the girl had been lifted to the honourable state of a maiden who has received a proposal.

In the holidays she came to her mother anxiously, and said: "Oh, mummy, I must turn Cyril [her brother] over to you, and you'll have to be very careful. He's read simply all Ibsen, and he's asking all sorts of questions, and I can't answer any more. Mind he doesn't trip *you* up."

It appeared that in those circles, male youth prefers the reading of plays to the reading of novels. I have heard the same news from publishers. I cannot imagine why the intelligently curious should prefer plays to novels, seeing that novels on the average are immensely superior in quality to plays. Still there you are. Facts are facts. And male youth does read novels, too. And in fiction it puts D. H. Lawrence first and the rest nowhere. It does not seem to read a lot of verse.

On the other hand, a middle-aged Yorkshire woollen-manufacturer asked my friend: " Can you tell me of any new poets? " An odd question from a Yorkshire woollen-manufacturer, thought my friend. " Why? " she inquired. The manufacturer replied: " There is a small group of us, and we meet of an evening to read poetry aloud to one another." Amazing? But why should it be amazing? Is there anything essentially strange in Yorkshire woollen-manufacturers having a poetical interest lively enough to make them read poetry aloud to one another? Even if it weren't quite natural, it ought to be. And it obviously was quite natural.

Personally I do not like to hear verse read aloud,—for the reason that it is so difficult to follow—much more difficult than prose, and so difficult to read aloud well. Nevertheless, when I am invited to give my predilections in the vast field of English verse, I usually read aloud Blake's " Songs of the Seasons " to the asker. I well know that I do not read enough verse, especially classical English verse. I wish I read more, but I seem seldom to be able to find the time. This is only an excuse, I admit. I don't read much verse because at heart I prefer to read prose. I used always to have by my bedside the works of Shakspeare and a Bible. I read the Bible but not Shakspeare. So in the end I put the Shakspeare volume back on a shelf. It struck me as pretentiously out of place by my bedside.

When I read classical poetry I do so for an ulterior purpose—because I find that if I am writing a novel or a story, the finest English verse has the capacity to lift me up out of the rut of composition and set me, and my work, on a higher plane. In other words, it inspires. This may not be the very best possible compliment to our poets, but it is a pretty good compliment all the same.

London.

Mrs. X, talking to me to-day in the street of the nursing-home where she had not long since undergone a most formidable operation, began by saying that it was a very good nursing-home. However, she soon modified her verdict and she ended by condemning the place totally. The following is a list of her objections to the place:

1. No hot and cold water supply in bedrooms.
2. No electric fires.
3. Very few gas-fires.
4. She had a smoking chimney in her room. This was apparently incurable. It was allowed to continue until the smoke gave her a bad throat. Then only was she removed to another room.
5. The noise of the nurse's starched aprons, which rustled at her slightest movement.
6. Noise of bells, night and day.

7. Noise of nurses chattering in corridors.
8. No certainty of freedom from invasion. No sure solitude.
9. Bad lunches. Bad teas.
10. Sameness and insufficiency of food.
11. No fruit. No salads. But an abundance of artificial laxatives.
12. Terrific and continuous piety of the matron. Matron and nurses were surprised that my friend had no religious books at her bedside.
13. The doctor was apparently financially interested in the Home. He actually said that nursing-homes only just paid their way.
14. No operating theatre in the Home. The doctor said that he preferred this. The operation was done in the patient's room, and the smell of ether, etc., lingered there for hours.
15. Fee, 18 guineas a week, with many extras. This fee, of course, did not include doctor's fees.
16. The nurses were overworked and underpaid —so far as she could gather.

London.

Some time ago the wife of the secretary of an association with which I am not unconnected wrote to me from a country town asking for an appointment to see me about a personal matter too confidential for a letter. I had never met her, and had met her husband only once—and that in a purely formal way. She came, by train and taxi, and the journey up, she said, had taken one and a half hours. A tall woman, with marked features, especially the nose, and a complexion whose crimson was not wholly due to rouge. I don't mean that she was Hebraic or that she drank. Neither. Fairly well dressed; new gloves, a gold bag. She crossed her legs sometimes; she never once sat straight with her back against the back of the chair, but as women used to sit when they wore bustles forty or fifty years ago. Aged 55 to 60. Energetic.

She said at once that her husband was underpaid, and that she wanted to earn money. Her bent was towards house-decoration. All she needed was a little capital, and would I lend her £150? She would positively pay it back within a year, with interest *if necessary*. She said she had come to me because of the personality revealed in my books. I had no intention of lending her any money at all, and I said almost at once that I couldn't. She said: "Don't refuse me." I told her that women-decorators were all over the place. She said she knew that; but she was also sure that *she* would succeed. She *knew* she could get work. She had had to refuse a large order a few months earlier because she lacked the capital to do it with. Etc. I asked her whether there was nobody else she could go to. No! I suggested a name. No! She couldn't go to him. She had comparatively rich relatives—her mother, among others—but it would be no use going to them. The fact was (she said) the richer people were the more calls were made upon them.

I tempered my refusal by the soft sympathy of my tone. I referred to my own obligations. I said, lying: "Believe me, if I could I would." She said: "I'm sure you would." She said her husband had no idea she was coming to me. She said: "Don't think, please, I'm asking you for a gift. Oh no! Of course I should pay it back. There would be *no* risk." And a lot more similar wild sayings. As for instance: "It isn't as if I was in a hole and came to you to get me out of it. No! This is for a legitimate enterprise." After about twenty minutes' talk she left.

She was a complete stranger to me. She had given me no hint in her letter of what she wanted. She had travelled some fifty miles to London, and she would (I suppose) have to travel some fifty miles home. And she thought I would finance her in the business of house-decorating, of which craft she had admittedly had no professional experience. There must be hundreds of women apparently sane who have tried to realize a dream similar to hers, but I should think few of them have had the nerve to go to a complete stranger for capital.

To-day, after quite a long interval I learnt that she made a regular practice of these begging expeditions, not one of which, according to the information given me, ever succeeded.

FOOTBALL MANNERS

London, December 12th.

If a set of young men from the East End or from some provincial centre of Association football had gone in mass formation to Twickenham Football ground last Monday and by force and rowdyism rendered impossible the playing of the inter-Varsity match, there would have been a loud outcry in the papers, and in all polite circles, against their ill-mannered lawlessness; the police-courts would have been densely populated next morning, and the non-payment of fines imposed would have ended in many doses of imprisonment.

Yet such conduct would have been no worse than the conduct, on that same day, of under-graduates from our ancient universities, which conduct began with processions on the tops of dining-tables in fashionable restaurants and ended in the breaking up of a performance in at least one West End theatre; and which conduct occupied only a few inches of space in the papers and was forgotten by an enlightened public in less than twenty-four hours. It was generally understood that University rowdyism in London had been finished for ever by certain outrageous, destructive antics last year. Not so.

The proof of the pudding is in the eating. If years of education at public schools and universities result in exhibitions of loutish violence which have no equal in Great Britain, what are we to think of the real value of such education? What-ever young men are taught at universities, they are not effectively taught either decency or good manners or self-control or self-respect for the elementary social rights of others. They are taught to behave like savages—and to be proud of it. The immediate cause of these disgraces is, of course, simple drunkenness, senseless and brutish indulgence in alcohol. The excuse offered for the youths is that they are young. Which plainly implies a theory that we ought not to expect citizens to be decent, civilized and law-abiding until they have reached the age of at least twenty-one. Is this a tenable theory?

London.

The Bible is full of mysticism, of which it is probably the finest treasury in existence, east or west. To my mind the most pregnant mystical exhortation ever written is: "Be still, and know that I am God." (Forty-sixth Psalm.) The first two words ought to be stressed and repeated thrice. The more one reads the Bible the more one perceives that it is permeated through and through with purely mystical emotion. Many religious people, and many readers of the Bible, seem to be insensible to mysticism, and are thus deprived of what is perhaps the deepest source of private comfort.

London.

I lunched with two statesmen at my political club. One of them told us a story about a man who had seen a suicide hanging. The man was asked by my first political friend: "But why didn't you cut him down?" The reply was: "Because he wasn't dead." My first political friend applied this story as a parable to the Tory Party. He said he didn't want to cut it down until it was dead. My second political friend agreed with much fervour. I remarked to both of them that they might have to wait quite some time. I did not go so far as to tell them that in my opinion the Tory Party will easily survive all other parties in this cautious and compromising country. All political parties in all countries disappear sooner or later, except the Conservative, and the Conservative is immortal because it is never for long divided against itself. How many times in Britain has the Liberal Party split? The first and most powerful instinct of Tories is self-preservation. They do not really want anything but the status quo. They are deeply aware that united they stand—not otherwise. And every Briton is at heart a Tory—especially every British Liberal.

Southern France.

One of the chief desirabilities of a holiday is that mechanical daily habits should be broken. If for instance you are accustomed, as many of us are, to the daily morning toil and grind and bore of physical exercises, deep breathing, and all the elaborate rigmarole of the toilet, you ought to be able, on an ideal holiday, to abolish the entire ceremonial. On an ideal holiday the house ought to be so warm, the climate so warm, and existence so informal that a suit of day pyjamas could be worn without anything else. To think of the sensation of freedom if you could bathe casually and carelessly, and live for the rest of the day in those pyjamas! Well, a woman could do this, and some occasionally do. But a man—he has to shave. Of course he absolutely need not shave; only he must. And shaving is enough to spoil the perfect idleness of any morning. No! The ideal holiday is impossible for men.

London.
A rich friend of mine gave a party to a number of fellow-clubmen. Not very late in the evening two facts became known. First that the supply of whiskey had run out. Second that the members of the hired orchestra were all drunk. So the party came to an end. Why it should have come to an end I could not understand. For neither whiskey nor music was really necessary to the continuation of a party of men who undoubtedly had considerable intellectual resources. Nevertheless, the party did come to an end.

London.
A woman friend of mine became very friendly with an excellent, genial manservant in a friend's house. One day he said to her, in a sudden access of confidence: "Madam, I wish to God you'd been born a housemaid."

Printed in Great Britain by
Butler & Tanner Ltd.,
Frome and London

F40 . 530